CW00487179

ISBN 978-1-333-38474-6
PIBN 10497953

FB &c Ltd, Dalton House, 60 Windsor Avenue, London, SW19 2RR.
Company number 08720141. Registered in England and Wales.

For support please visit www.forgottenbooks.com

UNIVERSITY OF CALIFORNIA PUBLICATIONS
IN
PHYSIOLOGY

Vol. 5, No. 4, pp. 23-69 March 29, 1919

TABLE OF PH, H^+ AND OH^- VALUES CORRESPONDING TO ELECTROMOTIVE FORCES DETERMINED IN HYDROGEN ELECTRODE MEASUREMENTS, WITH A BIBLIOGRAPHY.*

BY

CARL L. A. SCHMIDT AND D. R. HOAGLAND

(From the Hearst Laboratory of Pathology and Bacteriology, the Department of Biochemistry, and the Division of Agricultural Chemistry of the University of California.)

The importance of measurements of reaction or hydrogen ion concentration is well recognized in many fields of scientific work. In biological studies the reaction of the body tissues and fluids, optimum reaction for enzymes, and the dissociation of the proteins are among the many subjects investigated. Recent work has shown the importance of the reaction of the media for the proper growth of micro-organisms, and in the field of agricultural chemistry the relation of the acidity of the soil to the growth of the plant has also been indicated. The extensive application of the electrometric method for determining hydrogen ion concentration has led to its use in laboratories without extensive equipment, and where the extreme accuracy so necessary to the theoretical chemist is less essential than the ability quickly to carry out a large number of determinations without the use of refined apparatus. The results so obtained may still have an accuracy well within the limits of interpretation.

With a view of facilitating measurements of hydrogen ion concentration, one of us (Schmidt)[37] some years ago prepared tables for the conversion of voltages into hydrogen or hydroxyl ion concentrations, thus rendering unnecessary the somewhat tedious computations. Since

* Aided in part by a grant from the George Williams Hooper Foundation for Medical Research.

then Sörensen's suggestion of using P_H* as a measure of reaction instead of hydrogen ion concentration has been quite generally adopted and the expression of results simplified. For many purposes, however, both units are desirable. McClendon[172] has published a chart from which both of the above values within a certain range can be read. The continued demand for, as well as the advantages possessed by a direct conversion table led us to recalculate the former tables and include also the P_H values for both the normal and N/10 KCl-calomel electrode. For a range of voltages not frequently used the calculations have been made, as in the former tables, for every two millivolts, while for a certain range on either side of the neutral point this has been done for each millivolt.

We have also included an extensive bibliography, somewhat arbitrarily classified, since it is quite impossible to arrange the references according to the subjects treated without either duplicate citation or cross-indexing. While many of the references given possess merely historical interest, they have nevertheless been included and are of value as showing to what extent certain fields have been investigated. We have cited only those references which include determinations of hydrogen ion concentration by the electrometric method, by indicators, or by the use of buffer mixtures. Certain other references appertaining to the theory of the hydrogen electrode or methods of measuring electromotive forces have also been included.

Our calculations are based on the recent measurements by Lewis, Brighton, and Sebastian[61] of the potential of the normal and N/10 KCl-calomel electrodes and the dissociation constant of water. For the potential of the normal calomel electrode, referred to the potential of the normal hydrogen electrode as zero, we have taken

$$H_2,H^+ \ (M) \ || \ Hg, HgCl, KCl \ (M); \ E = 0.283 \text{ volt}$$

---

* The term PH is given to the exponent of 10 taken as a positive number. This is the most rational system since all values are expressed in the same units. Thus $C_H = 5.03 \times 10^{-10}$ can be expressed entirely as a power of 10.

$$5.03 = 10^{0.702} \ (\text{since } \log_{10} 5.03 = 0.702)$$
$$C_H = 10^{0.702} \times 10^{-10}$$
$$= 10^{-9.298}$$
$$P_H = 9.298$$

Another example: To find PH, when $C_H = 0.409 \times 10^{-7}$

$$P_H = \log_{10} \frac{1}{C_H}$$
$$P_H = \log_{10} 1 - \log_{10} C_H$$
$$\text{Log}_{10} C_H = 10^{-7} \times 10^{\bar{1}.612}$$
$$= \bar{8}.612$$
$$P_H = 17.388 - 10$$
$$P_H = 7.388$$

and for the difference in potential between the normal and N/10 KCl-calomel electrodes

$$Hg, HgCl, KCl\ (0.1\ M)\ \|\ HgCl, KCl, (M);\ E = 0.053.$$

This gives a value of 0.336 volt for the N/10 KCl-calomel electrode, a millivolt less than the value (0.337 volt) which most biological investigators have assigned to this electrode. However, if the latter value is adopted the table can still be used; it is merely necessary to shift the values given under the column $E_{\frac{N}{10}}$ one millivolt. For the dissociation constant of water we have taken

$$Kw = 1.012 \times 10^{-14}\ (25°C)$$

which gives a concentration of H^+ and OH^- as

$$\sqrt{Kw} = 1.006 \times 10^{-7}.$$

The value of Kw is stated by Lewis, Brighton, and Sebastian[61] to be correct within two or three per cent. All calculations in the table have been made on the temperature basis of 25°C. This value has been most generally used for physico-chemical work and is well within the range of room temperature in laboratories having no special facilities for temperature regulation. For small temperature variations this error, as will be shown later, is usually negligible.

For the calcuation of hydrogen ion concentration (C_H) the well-known Nernst equation

$$\pi = \frac{RT}{nF} \ln \frac{1}{(C_H)} \quad *$$

is used, where

R = gas constant in volt coulombs (8.31574).
T = absolute temperature (273.09 + 25).
n = valency of hydrogen (1).
F = Faraday constant (96,500 coulombs).
ln = natural logarithm. $\ln \frac{1}{C_H} = \frac{1}{0.4343} \log_{10} \frac{1}{C_H}$
π = difference of potential between the E.M.F. measured and the potential of the particular calomel cell used.
C_H = concentration of hydrogen ion to be determined.

At 25°C we have

$$\pi = 0.059152 \log_{10} \frac{1}{C_H} = 0.059152\ P_H.$$

* To be more exact, according to recent physico-chemical views it is the activity of the H-ion rather than the concentration which is measured. Numerically, however, the value is the same.

This equation gives at once both the values for P_H and C_H. To illustrate: when the voltage, using the N/10 KCl-calomel electrode, is 0.773

$$0.773 - 0.336 = 0.059152\ P_H = 0.059152 \log_{10} \frac{1}{C_H}$$

$$P_H = 7.388$$

$$P_H = \log_{10} \frac{1}{C_H}$$

$$C_H = \text{antilog.} \frac{1}{P_H}$$

$$C_H = 10.000 - 10$$

$$\frac{-7.388}{2.612 - 10} \text{ or } -7 \times \bar{1}.612$$

$$C_H = 0.409 \times 10^{-7}$$

$$C_{OH} = \frac{1.012 \times 10^{-14}}{0.409 \times 10^{-7}} = 2.47 \times 10^{-7}$$

The values calculated for C_H and P_H have been carried to three decimals and those for C_{OH} to two decimals. These are accurate to one unit in the last decimal place. For most purposes, however, it is sufficiently accurate to express values one decimal place less than given in the table.

The results given in the tables are based on a hydrogen pressure of 760 millimeters. It requires a considerable divergence from this pressure to produce a change in voltage within the accuracy of the tables. Loomis and Acree[65] have investigated the influence of pressure on the hydrogen electrode and found that a change of forty millimeters in the barometric pressure produced a change in the potential of only 0.0007 volt. It is evident that the ordinary barometric fluctuations are of no significance except in physico-chemical researches. If necessary, however, to correct for partial hydrogen pressures, it may be done as follows:

$$E_b = \text{E.M.F. measured at the barometric pressure b.}$$

$$e_b = E_b - 0.336 \text{ (for the N/ KCl-calomel electrode).}$$

$$e_{760} = e_b \pm \frac{0.05915}{2} \log \frac{760}{b}$$

$$E_{760} = e_{760} + 0.336.$$

The correction will be positive when b is less, and negative when greater than 760 millimeters.

Temperature has a somewhat greater influence on the potential of the hydrogen electrode, and varies with the range of C_{H^+} measured. Since our tables have been calculated on the temperature basis of

25°C, it will be necessary, if measurements are made at any other temperature, to convert the measured voltage to the value it would have at 25°C. Within ordinary temperature ranges the value for the calomel cell on the basis of $P_H = 0$ will not change, hence it is merely necessary to correct e_t.

E_t = voltage measured at the temperature t.
$e_t = E_t - 0.336$ (when the N/10 KCl-calomel electrode is used).
$e_{25} = e_t \times$ Factor.
E_{25} = voltage at 25°C.
$E_{25} = e_{25} + 0.336$.

From the Nernst equation it will be seen that the value for $\frac{RT}{nF}$ will change with the temperature, since T is the variable. We have calculated a series of factors for use within ordinary temperature ranges for the conversion of e_t to the value e_{25}.

Temperature	Factor (Multiply)
18	1.024
19	1.021
20	1.017
21	1.014
22	1.010
23	1.007
24	1.004
25	1.000
26	0.996
27	0.993
28	0.990
29	0.987
30	0.983

At the neutral point the change in voltage per degree of temperature variation will be about two millivolts, for $P_H = 4$ this will be only one millivolt, and for $P_H = 11$ it will rise to about three millivolts.

The large variety of apparatus used by different authors gives the investigator considerable choice for the particular purpose intended. For many biological, bacteriological, and agricultural investigations measurements accurate to two millivolts are within the limits of interpretation. For this purpose the method outlined by Hildebrand[27] and used extensively by Sharp and Hoagland,[365, 370] in which the voltage is directly measured by a voltmeter, is well adapted. The direct-reading potentiometer of Bovie[2] is useful for certain types of work. Where greater accuracy is required, such as the standardization of buffer mixtures, reaction changes occurring in the digestion of proteins by enzymes and the dissociation of proteins, the use of a potentiometer with a sensitive galvanometer is essential. The potentiometer manufactured by Leeds and Northrup is very convenient for this purpose.

Rapidity combined with accuracy is obtained by shaking the solution in contact with the hydrogen gas and electrode. For many purposes equilibrium may be quickly obtained by shaking the vessel by hand; more convenient, however, are the motor-driven shakers. For this purpose the electrode vessel designed by Clark[3] and used extensively by Clark and Lubs[111] is most convenient. Hydrogen is best generated electrolytically. Electrolysis of a 25 per cent KOH solution using nickel electrodes, or a 6 per cent H_2SO_4 solution using platinum electrodes and elimination of oxygen or ozone by passing the hydrogen over heated platinized asbestos, gives a very pure product. Compressed hydrogen from cylinders adequately purified may also be used. We have always used the Cottrell gauze electrode[118] since it is easily made and gives a large surface. Others have preferred an electrode made of a small sheet of platinum. This is coated with platinum black by deposition in a solution of H_2PtCl_6 containing a trace of lead acetate. Lewis, Brighton, and Sebastian[61] prefer an electrode of gold coated with iridium. For biological work it is essential that the electrode be saturated with hydrogen before immersing in the solution to be tested.

The best methods of evaluating the contact potential occurring at the junction of two liquids would be either by direct determination or by calculation, but since for most work neither of these methods can at present be used, it is necessary to reduce the contact potential to a minimum by interposing a saturated solution of KCl between the solution to be tested and the calomel electrode. For this purpose we use glass U tubes filled by placing in a heated solution of 2 per cent washed agar saturated with KCl. On cooling the agar solidifies and the KCl in part crystallizes. A fresh boundary, obtained by cutting small portions off the ends of the glass tube, should be used for each determination. Sand tubes, string dipped in a saturated KCl solution or a beaker containing the saturated KCl solution, into which an arm of the electrode vessel dips, diffusion being prevented by using ungreased stopcocks, have also been used. In measuring such systems as soil suspensions, certain protein solutions, etc., contamination from the KCl should be avoided by the use of a side arm or by leaving the agar tube in contact with the solution for a minimum time, otherwise interreactions may produce changes in the H^+ concentration. For most work the magnitude of the contact potential when a saturated solution of KCl is interposed is so small that it can be neglected. For very accurate work the extrapolation method of Bjerrum[88] gives the closest approximation.

Potential measurements of systems in which CO_2 is a determining factor in the reaction should be carried out in closed vessels, equilibrium being obtained by shaking. The methods used for determining the reaction of blood and body fluids are well adapted for this purpose. In soil extracts and nutrient solutions for plants an increase of alkalinity may result from the catalytic reduction of the NO_3 ion by the hydrogen gas. In solutions of high buffer value this effect does not result in any appreciable change in H-ion concentration, but in some solutions a serious error may be caused. It has been noted that thick coatings of platinum black have a much greater catalytic power than thin coatings; suitable precautions should be taken therefore when measuring solutions containing nitrates. A reduction of nitrates will be apparent in the gradual increase of voltage during the determination. Under these circumstances a constant value cannot be obtained.

There is very little choice between the normal and the N/10 KCl-calomel electrodes, since both are easily reproducible and remain constant. The saturated KCl-calomel electrode is less constant and has a greater temperature variation than either of the other calomel electrodes. Pure materials are necessary. For this purpose the methods described by Loomis and Acree[63] and Hildebrand[103] are well adapted. The glass vessels for the calomel electrode are designed to prevent contamination of the electrode solution and to provide a method for washing out the side arm with KCl solution. A convenient apparatus is described by Schmidt.[272]

For many purposes the use of indicators is a convenience. Many of the indicators used by Friedenthal,[398] Salm,[415] and others have largely been supplanted by indicators of the sulfonphthalein series,[404, 405] which offer more convenient ranges of color change. These have been extensively described by Clark and Lubs.[337] For some work it is necessary to test the applicability of a particular indicator by determination of the reaction by the electrometric method.

The use of buffer mixtures for the production of solutions containing definite concentrations of hydrogen ions has found extensive application. A good method whereby the accuracy of a hydrogen electrode system may be tested consists in the determination of the acidity or alkalinity of a carefully prepared buffer mixture. The range of an indicator may likewise be tested this way. Buffer mixtures are usually solutions of acetates, bicarbonates, borates, phosphates, phthalates, or cacodylates. These have been carefully standardized and described in the work of Sörensen,[38] Palitzsch,[113] Clark[111] and others.

TABLES

$E_{\frac{N}{1}}$	$E_{\frac{N}{10}}$	P_H	C_{H^+} $\times$ N H^+	C_{OH^-} $\times 10^{-14}$ OH^-
0.283	0.336	....	1.000	1.01
0.285	0.338	0.034	0.925	1.09
0.287	0.340	0.068	0.856	1.18
0.289	0.342	0.101	0.792	1.28
0.291	0.344	0.135	0.732	1.38
0.293	0.346	0.169	0.678	1.49
0.295	0.348	0.203	0.627	1.61
0.297	0.350	0.237	0.580	1.75
0.299	0.352	0.270	0.536	1.89
0.301	0.354	0.304	0.496	2.04
0.303	0.356	0.338	0.459	2.20
0.305	0.358	0.372	0.425	2.38
0.307	0.360	0.406	0.393	2.58
0.309	0.362	0.440	0.364	2.78
0.311	0.364	0.473	0.336	3.01
0.313	0.366	0.507	0.311	3.25
0.315	0.368	0.541	0.288	3.51
0.317	0.370	0.575	0.266	3.80
0.319	0.372	0.609	0.246	4.11
0.321	0.374	0.642	0.228	4.44
0.323	0.376	0.676	0.211	4.80
0.325	0.378	0.710	0.195	5.19
0.327	0.380	0.744	0.180	5.62
0.329	0.382	0.778	0.167	6.06
0.331	0.384	0.811	0.154	6.57
0.333	0.386	0.845	0.143	7.08
0.335	0.388	0.879	0.132	7.67
0.337	0.390	0.913	0.122	8.30
0.339	0.392	0.947	0.113	8.96
0.341	0.394	0.980	0.105	9.64

$E_{\frac{N}{1}}$	$E_{\frac{N}{10}}$	P_H	C_{H^+} $\times 10^{-1}$ H^+	C_{OH^-} $\times 10^{-13}$ OH^-
0.343	0.396	1.014	0.968	1.05
0.345	0.398	1.048	0.895	1.13
0.347	0.400	1.082	0.828	1.22
0.349	0.402	1.116	0.766	1.32
0.351	0.404	1.150	0.709	1.43
0.353	0.406	1.183	0.656	1.54
0.355	0.408	1.217	0.607	1.67
0.357	0.410	1.251	0.561	1.80
0.359	0.412	1.285	0.519	1.95
0.361	0.414	1.319	0.480	2.11
0.363	0.416	1.352	0.444	2.28
0.365	0.418	1.386	0.411	2.46
0.367	0.420	1.420	0.380	2.66

$E_{N/1}$	$E_{N/10}$	P_H	C_{H^+} $\times 10^{-1}$ H^+	C_{OH^-} $\times 10^{-13}$ OH^-
0.369	0.422	1.454	0.352	2.88
0.371	0.424	1.488	0.325	3.11
0.373	0.426	1.521	0.301	3.36
0.375	0.428	1.555	0.278	3.64
0.377	0.430	1.589	0.258	3.92
0.379	0.432	1.623	0.238	4.25
0.381	0.434	1.657	0.221	4.58
0.383	0.436	1.691	0.204	4.96
0.385	0.438	1.724	0.189	5.35
0.387	0.440	1.758	0.175	5.78
0.389	0.442	1.792	0.162	6.25
0.391	0.444	1.826	0.149	6.79
0.393	0.446	1.860	0.138	7.33
0.395	0.448	1.893	0.128	7.91
0.397	0.450	1.927	0.118	8.58
0.399	0.452	1.961	0.109	9.28
0.401	0.454	1.995	0.101	10.00

$E_{N/1}$	$E_{N/10}$	P_H	C_{H^+} $\times 10^{-2}$ H^+	C_{OH^-} $\times 10^{-12}$ OH^-
0.403	0.456	2.029	0.936	1.08
0.405	0.458	2.062	0.866	1.17
0.407	0.460	2.096	0.801	1.26
0.409	0.462	2.130	0.741	1.37
0.411	0.464	2.164	0.686	1.48
0.413	0.466	2.198	0.634	1.60
0.415	0.468	2.232	0.587	1.72
0.417	0.470	2.265	0.543	1.86
0.419	0.472	2.299	0.502	2.02
0.421	0.474	2.333	0.465	2.18
0.423	0.476	2.367	0.430	2.35
0.425	0.478	2.401	0.398	2.54
0.427	0.480	2.434	0.368	2.75
0.429	0.482	2.468	0.340	2.98
0.431	0.484	2.502	0.315	3.21
0.433	0.486	2.536	0.291	3.48
0.435	0.488	2.570	0.269	3.76
0.437	0.490	2.603	0.249	4.06
0.439	0.492	2.637	0.231	4.38
0.441	0.494	2.671	0.213	4.75
0.443	0.496	2.705	0.197	5.14
0.445	0.498	2.739	0.183	5.53
0.447	0.500	2.772	0.169	5.99
0.449	0.502	2.806	0.156	6.49
0.451	0.504	2.840	0.145	6.98
0.453	0.506	2.874	0.134	7.55
0.455	0.508	2.908	0.124	8.16
0.457	0.510	2.942	0.114	8.88
0.459	0.512	2.975	0.106	9.55

$E_{\frac{N}{1}}$	$E_{\frac{N}{10}}$	P_H	C_{H^+} $\times 10^{-3}$ H^+	C_{OH^-} $\times 10^{-11}$ OH^-
0.461	0.514	3.009	0.979	1.03
0.463	0.516	3.043	0.906	1.12
0.465	0.518	3.077	0.838	1.21
0.467	0.520	3.111	0.775	1.31
0.469	0.522	3.144	0.717	1.41
0.471	0.524	3.178	0.663	1.53
0.473	0.526	3.212	0.614	1.65
0.475	0.528	3.246	0.568	1.78
0.477	0.530	3.280	0.525	1.93
0.479	0.532	3.313	0.486	2.08
0.481	0.534	3.347	0.450	2.25
0.483	0.536	3.381	0.416	2.43
0.485	0.538	3.415	0.385	2.63
0.487	0.540	3.449	0.356	2.84
0.489	0.542	3.483	0.329	3.08
0.491	0.544	3.516	0.305	3.32
0.493	0.546	3.550	0.282	3.59
0.495	0.548	3.584	0.261	3.88
0.497	0.550	3.618	0.241	4.20
0.499	0.552	3.652	0.223	4.54
0.501	0.554	3.685	0.206	4.91
0.503	0.556	3.719	0.191	5.30
0.505	0.558	3.753	0.177	5.72
0.507	0.560	3.787	0.163	6.21
0.509	0.562	3.821	0.151	6.70
0.511	0.564	3.854	0.140	7.23
0.513	0.566	3.888	0.129	7.85
0.515	0.568	3.922	0.120	8.43
0.517	0.570	3.956	0.111	9.12
0.519	0.572	3.990	0.102	9.92

$E_{\frac{N}{1}}$	$E_{\frac{N}{10}}$	P_H	C_{H^+} $\times 10^{-4}$ H^+	C_{OH^-} $\times 10^{-10}$ OH^-
0.521	0.574	4.023	0.947	1.07
0.522	0.575	4.040	0.911	1.11
0.523	0.576	4.057	0.876	1.16
0.524	0.577	4.074	0.843	1.20
0.525	0.578	4.091	0.811	1.25
0.526	0.579	4.108	0.780	1.30
0.527	0.580	4.125	0.750	1.35
0.528	0.581	4.142	0.721	1.40
0.529	0.582	4.159	0.694	1.46
0.530	0.583	4.176	0.667	1.52
0.531	0.584	4.193	0.642	1.58
0.532	0.585	4.210	0.617	1.64
0.533	0.586	4.226	0.594	1.70
0.534	0.587	4.243	0.571	1.77
0.535	0.588	4.260	0.549	1.84
0.536	0.589	4.277	0.528	1.92
0.537	0.590	4.294	0.508	1.99
0.538	0.591	4.311	0.489	2.07

E_N / 1	E_N / 10	P_H	C_{H^+} $\times 10^{-4}$ H^+	C_{OH^-} $\times 10^{-10}$ OH^-
0.539	0.592	4.328	0.470	2.15
0.540	0.593	4.345	0.452	2.24
0.541	0.594	4.362	0.435	2.33
0.542	0.595	4.379	0.418	2.42
0.543	0.596	4.395	0.402	2.52
0.544	0.597	4.412	0.387	2.61
0.545	0.598	4.429	0.372	2.72
0.546	0.599	4.446	0.358	2.83
0.547	0.600	4.463	0.344	2.94
0.548	0.601	4.480	0.331	3.06
0.549	0.602	4.497	0.319	3.17
0.550	0.603	4.514	0.306	3.31
0.551	0.604	4.531	0.295	3.43
0.552	0.605	4.548	0.283	3.58
0.553	0.606	4.564	0.273	3.71
0.554	0.607	4.581	0.262	3.86
0.555	0.608	4.598	0.252	4.02
0.556	0.609	4.615	0.243	4.16
0.557	0.610	4.632	0.233	4.34
0.558	0.611	4.649	0.224	4.52
0.559	0.612	4.666	0.216	4.69
0.560	0.613	4.683	0.208	4.87
0.561	0.614	4.700	0.200	5.06
0.562	0.615	4.717	0.192	5.27
0.563	0.616	4.734	0.185	5.47
0.564	0.617	4.750	0.178	5.69
0.565	0.618	4.767	0.171	5.92
0.566	0.619	4.784	0.164	6.17
0.567	0.620	4.801	0.158	6.41
0.568	0.621	4.818	0.152	6.66
0.569	0.622	4.835	0.146	6.93
0.570	0.623	4.852	0.141	7.18
0.571	0.624	4.869	0.135	7.50
0.572	0.625	4.886	0.130	7.78
0.573	0.626	4.903	0.125	8.10
0.574	0.627	4.920	0.120	8.43
0.575	0.628	4.936	0.116	8.72
0.576	0.629	4.953	0.111	9.12
0.577	0.630	4.970	0.107	9.46
0.578	0.631	4.987	0.103	9.83

E_N / 1	E_N / 10	P_H	C_{H^+} $\times 10^{-5}$ H^+	C_{OH^-} $\times 10^{-9}$ OH^-
0.579	0.632	5.004	0.991	1.02
0.580	0.633	5.021	0.953	1.06
0.581	0.634	5.038	0.916	1.10
0.582	0.635	5.055	0.881	1.15
0.583	0.636	5.072	0.848	1.19
0.584	0.637	5.089	0.815	1.24
0.585	0.638	5.106	0.784	1.29
0.586	0.639	5.122	0.754	1.34

0.587	0.640	5.139	0.725	1.40
0.588	0.641	5.156	0.698	1.45
0.589	0.642	5.173	0.671	1.51
0.590	0.643	5.190	0.646	1.57
0.591	0.644	5.207	0.621	1.63
0.592	0.645	5.224	0.597	1.70
0.593	0.646	5.241	0.574	1.76
0.594	0.647	5.258	0.552	1.83
0.595	0.648	5.275	0.531	1.91
0.596	0.649	5.292	0.511	1.98
0.597	0.650	5.308	0.492	2.06
0.598	0.651	5.325	0.473	2.14
0.599	0.652	5.342	0.455	2.22
0.600	0.653	5.359	0.437	2.32
0.601	0.654	5.376	0.421	2.40
0.602	0.655	5.393	0.405	2.50
0.603	0.656	5.410	0.390	2.59
0.604	0.657	5.427	0.374	2.71
0.605	0.658	5.444	0.360	2.81
0.606	0.659	5.461	0.346	2.92
0.607	0.660	5.478	0.333	3.04
0.608	0.661	5.495	0.320	3.16
0.609	0.662	5.511	0.308	3.29
0.610	0.663	5.528	0.296	3.42
0.611	0.664	5.545	0.285	3.55
0.612	0.665	5.562	0.274	3.69
0.613	0.666	5.579	0.264	3.83
0.614	0.667	5.596	0.254	3.98
0.615	0.668	5.613	0.244	4.15
0.616	0.669	5.630	0.235	4.31
0.617	0.670	5.647	0.226	4.48
0.618	0.671	5.664	0.217	4.66
0.619	0.672	5.681	0.209	4.84
0.620	0.673	5.697	0.201	5.03
0.621	0.674	5.714	0.193	5.24
0.622	0.675	5.731	0.186	5.44
0.623	0.676	5.748	0.179	5.65
0.624	0.677	5.765	0.172	5.88
0.625	0.678	5.782	0.165	6.13
0.626	0.679	5.799	0.159	6.36
0.627	0.680	5.816	0.153	6.61
0.628	0.681	5.833	0.147	6.88
0.629	0.682	5.850	0.141	7.18
0.630	0.683	5.866	0.136	7.44
0.631	0.684	5.883	0.131	7.73
0.632	0.685	5.900	0.126	8.03
0.633	0.686	5.917	0.121	8.36
0.634	0.687	5.934	0.116	8.72
0.635	0.688	5.951	0.112	9.04
0.636	0.689	5.968	0.108	9.37
0.637	0.690	5.985	0.104	9.73

E_N / $\frac{N}{1}$	E_N / $\frac{N}{10}$	P_H	C_{H^+} $\times 10^{-6}$ H^+	C_{OH^-} $\times 10^{-8}$ OH^-
0.638	0.691	6.002	0.996	1.02
0.639	0.692	6.019	0.958	1.06
0.640	0.693	6.036	0.921	1.10
0.641	0.694	6.052	0.886	1.14
0.642	0.695	6.069	0.852	1.19
0.643	0.696	6.086	0.820	1.23
0.644	0.697	6.103	0.789	1.28
0.645	0.698	6.120	0.758	1.34
0.646	0.699	6.137	0.729	1.39
0.647	0.700	6.154	0.702	1.44
0.648	0.701	6.171	0.675	1.50
0.649	0.702	6.188	0.649	1.56
0.650	0.703	6.204	0.625	1.62
0.651	0.704	6.221	0.601	1.68
0.652	0.705	6.238	0.578	1.75
0.653	0.706	6.255	0.556	1.82
0.654	0.707	6.272	0.535	1.89
0.655	0.708	6.289	0.514	1.97
0.656	0.709	6.306	0.495	2.04
0.657	0.710	6.323	0.476	2.13
0.658	0.711	6.340	0.458	2.21
0.659	0.712	6.357	0.440	2.30
0.660	0.713	6.373	0.423	2.39
0.661	0.714	6.390	0.407	2.49
0.662	0.715	6.407	0.392	2.58
0.663	0.716	6.424	0.377	2.68
0.664	0.717	6.441	0.362	2.80
0.665	0.718	6.458	0.348	2.91
0.666	0.719	6.475	0.335	3.02
0.667	0.720	6.492	0.322	3.14
0.668	0.721	6.509	0.310	3.26
0.669	0.722	6.526	0.298	3.40
0.670	0.723	6.543	0.287	3.53
0.671	0.724	6.559	0.276	3.67
0.672	0.725	6.576	0.265	3.82
0.673	0.726	6.593	0.255	3.97
0.674	0.727	6.610	0.245	4.13
0.675	0.728	6.627	0.236	4.29
0.676	0.729	6.644	0.227	4.46
0.677	0.730	6.661	0.218	4.64
0.678	0.731	6.678	0.210	4.82
0.679	0.732	6.695	0.202	5.01
0.680	0.733	6.712	0.194	5.22
0.681	0.734	6.728	0.187	5.41
0.682	0.735	6.745	0.180	5.62
0.683	0.736	6.762	0.173	5.85
0.684	0.737	6.779	0.166	6.10
0.685	0.738	6.796	0.160	6.32
0.686	0.739	6.813	0.154	6.57
0.687	0.740	6.830	0.148	6.84

$\overline{1}$	$\overline{10}$		$\times 10^{-6}$ H^+	$\times 10^{-8}$ OH^-
0.688	0.741	6.847	0.142	7.13
0.689	0.742	6.864	0.137	7.39
0.690	0.743	6.881	0.132	7.67
0.691	0.744	6.898	0.127	7.97
0.692	0.745	6.914	0.122	8.30
0.693	0.746	6.931	0.117	8.65
0.694	0.747	6.948	0.113	8.96
0.695	0.748	6.965	0.108	9.37
0.696	0.749	6.982	0.104	9.73
*0.697	0.750	6.999	0.100	10.12

E_N $\overline{1}$	E_N $\overline{10}$	P_H	C_{H^+} $\times 10^{-7}$ H^+	C_{OH^-} $\times 10^{-7}$ OH^-
0.698	0.751	7.016	0.964	1.05
0.699	0.752	7.033	0.927	1.09
0.700	0.753	7.050	0.892	1.13
0.701	0.754	7.067	0.858	1.18
0.702	0.755	7.084	0.825	1.23
0.703	0.756	7.100	0.794	1.27
0.704	0.757	7.117	0.763	1.33
0.705	0.758	7.134	0.734	1.38
0.706	0.759	7.151	0.706	1.43
0.707	0.760	7.168	0.679	1.49
0.708	0.761	7.185	0.653	1.55
0.709	0.762	7.202	0.628	1.61
0.710	0.763	7.219	0.604	1.68
0.711	0.764	7.236	0.581	1.74
0.712	0.765	7.253	0.559	1.81
0.713	0.766	7.269	0.538	1.88
0.714	0.767	7.286	0.517	1.96
0.715	0.768	7.303	0.497	2.04
0.716	0.769	7.320	0.478	2.12
0.717	0.770	7.337	0.460	2.20
0.718	0.771	7.354	0.443	2.28
0.719	0.772	7.371	0.426	2.38
0.720	0.773	7.388	0.409	2.47
0.721	0.774	7.405	0.394	2.57
0.722	0.775	7.422	0.379	2.67
0.723	0.776	7.439	0.364	2.78
0.724	0.777	7.455	0.350	2.89
0.725	0.778	7.472	0.337	3.00
0.726	0.779	7.489	0.324	3.12
0.727	0.780	7.506	0.312	3.24
0.728	0.781	7.523	0.300	3.37
0.729	0.782	7.540	0.288	3.51
0.730	0.783	7.557	0.277	3.65
0.731	0.784	7.574	0.267	3.79
0.732	0.785	7.591	0.257	3.94
0.733	0.786	7.608	0.247	4.10

* Neutral point.

E_N $\overline{1}$	E_N $\overline{10}$	PH	C_{H^+} $\times 10^{-7}$ H^+	C_{OH^-} $\times 10^{-7}$ OH^-
0.734	0.787	7.624	0.238	4.25
0.735	0.788	7.641	0.228	4.44
0.736	0.789	7.658	0.220	4.60
0.737	0.790	7.675	0.211	4.80
0.738	0.791	7.692	0.203	4.99
0.739	0.792	7.709	0.195	5.19
0.740	0.793	7.726	0.188	5.38
0.741	0.794	7.743	0.181	5.59
0.742	0.795	7.760	0.174	5.82
0.743	0.796	7.777	0.167	6.06
0.744	0.797	7.794	0.161	6.29
0.745	0.798	7.810	0.155	6.53
0.746	0.799	7.827	0.149	6.79
0.747	0.800	7.844	0.143	7.08
0.748	0.801	7.861	0.138	7.33
0.749	0.802	7.878	0.132	7.67
0.750	0.803	7.895	0.127	7.97
0.751	0.804	7.912	0.123	8.23
0.752	0.805	7.929	0.118	8.58
0.753	0.806	7.946	0.113	8.96
0.754	0.807	7.963	0.109	9.28
0.755	0.808	7.980	0.105	9.64
0.756	0.809	7.996	0.101	10.02

E_N $\overline{1}$	E_N $\overline{10}$	PH	C_{H^+} $\times 10^{-8}$ H^+	C_{OH^-} $\times 10^{-6}$ OH^-
0.757	0.810	8.013	0.970	1.04
0.758	0.811	8.030	0.933	1.08
0.759	0.812	8.047	0.897	1.13
0.760	0.813	8.064	0.863	1.17
0.761	0.814	8.081	0.830	1.22
0.762	0.815	8.098	0.798	1.27
0.763	0.816	8.115	0.768	1.32
0.764	0.817	8.132	0.739	1.37
0.765	0.818	8.149	0.710	1.43
0.766	0.819	8.165	0.683	1.48
0.767	0.820	8.182	0.657	1.54
0.768	0.821	8.199	0.632	1.60
0.769	0.822	8.216	0.608	1.66
0.770	0.823	8.233	0.585	1.73
0.771	0.824	8.250	0.562	1.80
0.772	0.825	8.267	0.541	1.87
0.773	0.826	8.284	0.520	1.95
0.774	0.827	8.301	0.500	2.02
0.775	0.828	8.318	0.481	2.10
0.776	0.829	8.335	0.463	2.19
0.777	0.830	8.351	0.445	2.27
0.778	0.831	8.368	0.428	2.36
0.779	0.832	8.385	0.412	2.46

0.780	0.833	8.402	0.396	2.56
0.781	0.834	8.419	0.381	2.66
0.782	0.835	8.436	0.367	2.76
0.783	0.836	8.453	0.353	2.87
0.784	0.837	8.470	0.339	2.99
0.785	0.838	8.487	0.326	3.10
0.786	0.839	8.504	0.314	3.22
0.787	0.840	8.521	0.302	3.35
0.788	0.841	8.537	0.290	3.49
0.789	0.842	8.554	0.279	3.63
0.790	0.843	8.571	0.269	3.76
0.791	0.844	8.588	0.258	3.92
0.792	0.845	8.605	0.248	4.08
0.793	0.846	8.622	0.239	4.23
0.794	0.847	8.639	0.230	4.40
0.795	0.848	8.656	0.221	4.58
0.796	0.849	8.673	0.213	4.75
0.797	0.850	8.690	0.204	4.96
0.798	0.851	8.706	0.197	5.14
0.799	0.852	8.723	0.189	5.35
0.800	0.853	8.740	0.182	5.56
0.801	0.854	8.757	0.175	5.78
0.802	0.855	8.774	0.168	6.02
0.803	0.856	8.791	0.162	6.25
0.804	0.857	8.808	0.156	6.49
0.805	0.858	8.825	0.150	6.75
0.806	0.859	8.842	0.144	7.03
0.807	0.860	8.859	0.139	7.28
0.808	0.861	8.876	0.133	7.61
0.809	0.862	8.892	0.128	7.91
0.810	0.863	8.909	0.123	8.23
0.811	0.864	8.926	0.119	8.50
0.812	0.865	8.943	0.114	8.88
0.813	0.866	8.960	0.110	9.20
0.814	0.867	8.977	0.106	9.55
0.815	0.868	8.994	0.101	10.00

E_N / $\frac{}{1}$	E_N / $\frac{}{10}$	P_H	C_{H^+} $\times 10^{-9}$ H^+	C_{OH^-} $\times 10^{-5}$ OH^-
0.816	0.869	9.011	0.975	1.04
0.817	0.870	9.028	0.938	1.08
0.818	0.871	9.045	0.902	1.12
0.819	0.872	9.062	0.868	1.17
0.820	0.873	9.078	0.835	1.21
0.821	0.874	9.095	0.803	1.26
0.822	0.875	9.112	0.772	1.31
0.823	0.876	9.129	0.743	1.36
0.824	0.877	9.146	0.714	1.42
0.825	0.878	9.163	0.687	1.47

$\frac{E_N}{1}$	$\frac{E_N}{10}$	P_H	C_{H^+} $\times 10^{-9}$ H^+	C_{OH^-} $\times 10^{-5}$ OH^-
0.826	0.879	9.180	0.661	1.53
0.827	0.880	9.197	0.636	1.59
0.828	0.881	9.214	0.611	1.66
0.829	0.882	9.231	0.588	1.72
0.830	0.883	9.248	0.566	1.79
0.831	0.884	9.264	0.544	1.86
0.832	0.885	9.281	0.523	1.93
0.833	0.886	9.298	0.503	2.01
0.834	0.887	9.315	0.484	2.09
0.835	0.888	9.332	0.466	2.17
0.836	0.889	9.349	0.448	2.26
0.837	0.890	9.366	0.431	2.35
0.838	0.891	9.383	0.414	2.44
0.839	0.892	9.400	0.398	2.54
0.840	0.893	9.417	0.383	2.64
0.841	0.894	9.434	0.369	2.74
0.842	0.895	9.450	0.354	2.86
0.843	0.896	9.467	0.341	2.97
0.844	0.897	9.484	0.328	3.09
0.845	0.898	9.501	0.315	3.21
0.846	0.899	9.518	0.304	3.33
0.847	0.900	9.535	0.292	3.47
0.848	0.901	9.552	0.281	3.60
0.849	0.902	9.569	0.270	3.75
0.850	0.903	9.585	0.260	3.89
0.851	0.904	9.602	0.250	4.05
0.852	0.905	9.619	0.240	4.22
0.853	0.906	9.636	0.231	4.38
0.854	0.907	9.653	0.222	4.56
0.855	0.908	9.670	0.214	4.73
0.856	0.909	9.687	0.206	4.91
0.857	0.910	9.704	0.198	5.11
0.858	0.911	9.721	0.190	5.33
0.859	0.912	9.738	0.183	5.53
0.860	0.913	9.755	0.176	5.75
0.861	0.914	9.772	0.169	5.99
0.862	0.915	9.788	0.163	6.21
0.863	0.916	9.805	0.157	6.45
0.364	0.917	9.822	0.151	6.70
0.865	0.918	9.839	0.145	6.98
0.866	0.919	9.856	0.139	7.28
0.867	0.920	9.873	0.134	7.55
0.868	0.921	9.890	0.129	7.84
0.869	0.922	9.907	0.124	8.16
0.870	0.923	9.924	0.119	8.50
0.871	0.924	9.941	0.115	8.80
0.872	0.925	9.958	0.110	9.20
0.873	0.926	9.974	0.106	9.55
0.874	0.927	9.991	0.102	9.92

E_N $\frac{}{1}$	E_N $\frac{}{10}$	P_H	C_{H^+} $\times 10^{-10}$ H^+	C_{OH^-} $\times 10^{-4}$ OH^-
0.875	0.928	10.008	0.981	1.03
0.877	0.930	10.042	0.908	1.11
0.879	0.932	10.076	0.840	1.20
0.881	0.934	10.110	0.777	1.30
0.883	0.936	10.143	0.719	1.41
0.885	0.938	10.177	0.665	1.52
0.887	0.940	10.211	0.615	1.65
0.889	0.942	10.245	0.569	1.78
0.891	0.944	10.279	0.526	1.92
0.893	0.946	10.313	0.487	2.08
0.895	0.948	10.346	0.451	2.24
0.897	0.950	10.380	0.417	2.43
0.899	0.952	10.414	0.386	2.62
0.901	0.954	10.448	0.357	2.83
0.903	0.956	10.481	0.330	3.07
0.905	0.958	10.515	0.305	3.32
0.907	0.960	10.549	0.282	3.59
0.909	0.962	10.583	0.261	3.88
0.911	0.964	10.617	0.242	4.18
0.913	0.966	10.651	0.224	4.52
0.915	0.968	10.684	0.207	4.89
0.917	0.970	10.718	0.191	5.30
0.919	0.972	10.752	0.177	5.72
0.921	0.974	10.786	0.164	6.17
0.923	0.976	10.820	0.152	6.66
0.925	0.978	10.853	0.140	7.23
0.927	0.980	10.887	0.130	7.78
0.929	0.982	10.921	0.120	8.43
0.931	0.984	10.955	0.111	9.12
0.933	0.986	10.989	0.103	9.83

E_N $\frac{}{1}$	E_N $\frac{}{10}$	P_H	C_{H^+} $\times 10^{-11}$ H^+	C_{OH^-} $\times 10^{-3}$ OH^-
0.935	0.988	11.022	0.950	1.07
0.937	0.990	11.056	0.879	1.15
0.939	0.992	11.090	0.813	1.24
0.941	0.994	11.124	0.752	1.35
0.943	0.996	11.158	0.696	1.45
0.945	0.998	11.191	0.644	1.57
0.947	1.000	11.225	0.595	1.70
0.949	1.002	11.259	0.551	1.84
0.951	1.004	11.293	0.509	1.99
0.953	1.006	11.327	0.471	2.15
0.955	1.008	11.361	0.436	2.32
0.957	1.010	11.394	0.403	2.51
0.959	1.012	11.428	0.373	2.71
0.961	1.014	11.462	0.345	2.93
0.963	1.016	11.496	0.319	3.17
0.965	1.018	11.530	0.295	3.43

$E_{\frac{N}{1}}$	$E_{\frac{N}{10}}$	P_H	C_{H^+} $\times 10^{-11}$ H^+	C_{OH^-} $\times 10^{-3}$ OH^-
0.967	1.020	11.563	0.273	3.71
0.969	1.022	11.597	0.253	4.00
0.971	1.024	11.631	0.234	4.32
0.973	1.026	11.665	0.216	4.69
0.975	1.028	11.699	0.200	5.06
0.977	1.030	11.732	0.185	5.47
0.979	1.032	11.766	0.171	5.92
0.981	1.034	11.800	0.159	6.36
0.983	1.036	11.834	0.147	6.88
0.985	1.038	11.868	0.136	7.44
0.987	1.040	11.901	0.126	8.03
0.989	1.042	11.935	0.116	8.72
0.991	1.044	11.969	0.107	9.46

$E_{\frac{N}{1}}$	$E_{\frac{N}{10}}$	P_H	C_{H^+} $\times 10^{-12}$ H^+	C_{OH^-} $\times 10^{-2}$ OH^-
0.993	1.046	12.003	0.993	1.02
0.995	1.048	12.037	0.919	1.10
0.997	1.050	12.071	0.850	1.19
0.999	1.052	12.104	0.786	1.29
1.001	1.054	12.138	0.728	1.39
1.003	1.056	12.172	0.673	1.50
1.005	1.058	12.206	0.623	1.62
1.007	1.060	12.240	0.576	1.76
1.009	1.062	12.273	0.533	1.90
1.011	1.064	12.307	0.493	2.05
1.013	1.066	12.341	0.456	2.22
1.015	1.068	12.375	0.422	2.40
1.017	1.070	12.409	0.390	2.59
1.019	1.072	12.443	0.361	2.80
1.021	1.074	12.476	0.334	3.03
1.023	1.076	12.510	0.309	3.28
1.025	1.078	12.544	0.286	3.54
1.027	1.080	12.578	0.264	3.83
1.029	1.082	12.612	0.245	4.13
1.031	1.084	12.645	0.226	4.48
1.033	1.086	12.679	0.209	4.84
1.035	1.088	12.713	0.194	5.22
1.037	1.090	12.747	0.179	5.65
1.039	1.092	12.781	0.166	6.10
1.041	1.094	12.814	0.153	6.61
1.043	1.096	12.848	0.142	7.13
1.045	1.098	12.882	0.131	7.73
1.047	1.100	12.916	0.121	8.36
1.049	1.102	12.950	0.112	9.04
1.051	1.104	12.983	0.104	9.73

$E_{\frac{N}{1}}$	$E_{\frac{N}{10}}$	P_H	C_{H^+} $\times 10^{-13}$ H^+	C_{OH^-} $\times 10^{-1}$ OH^-
1.053	1.106	13.017	0.961	1.05
1.055	1.108	13.051	0.889	1.14
1.057	1.110	13.085	0.822	1.23
1.059	1.112	13.119	0.761	1.33
1.061	1.114	13.153	0.704	1.44
1.063	1.116	13.186	0.651	1.55
1.065	1.118	13.220	0.602	1.68
1.067	1.120	13.254	0.557	1.82
1.069	1.122	13.288	0.516	1.96
1.071	1.124	13.322	0.477	2.12
1.073	1.126	13.355	0.441	2.29
1.075	1.128	13.389	0.408	2.48
1.077	1.130	13.423	0.378	2.68
1.079	1.132	13.457	0.349	2.90
1.081	1.134	13.491	0.323	3.13
1.083	1.136	13.524	0.299	3.38
1.085	1.138	13.558	0.277	3.65
1.087	1.140	13.592	0.256	3.95
1.089	1.142	13.626	0.237	4.27
1.091	1.144	13.660	0.219	4.62
1.093	1.146	13.693	0.203	4.99
1.095	1.148	13.727	0.187	5.41
1.097	1.150	13.761	0.173	5.85
1.099	1.152	13.795	0.160	6.32
1.101	1.154	13.829	0.148	6.84
1.103	1.156	13.863	0.137	7.39
1.105	1.158	13.896	0.127	7.97
1.107	1.160	13.930	0.117	8.65
1.109	1.162	13.964	0.109	9.28
1.111	1.164	13.998	0.101	10.02
			$\times 10^{-14}$ H^+	$\times$ N OH^-
1.113	1.166	14.032	0.930	1.09

LITERATURE

I. GENERAL AND THEORETICAL.

A. APPARATUS.

[1] Barendrecht, H. P., A simple hydrogen electrode, Biochem. Jour., 1915, vol. 9, pp. 66–70.

[2] Bovie, W. T., A direct reading potentiometer for measuring and recording both the actual and the total reaction of solutions, Jour. Med. Res., 1915, vol. 33, pp. 295–322.

[3] Clark, W. M., A hydrogen electrode vessel, Jour. Biol. Chem., 1915, vol. 23, pp. 475–486.

[4] Liebermann, L. v., Platinelektroden zur Bestimmung der H– und OH– Ionenkonzentration, Chem. Ztg., 1911, vol. 35, p. 972.

[5] Long, J. H., A simple cell for the determination of hydrogen ion concentration, Jour. Am. Chem. Soc., 1916, vol. 38, pp. 936–939.

[6] McClendon, J. F., New hydrogen electrodes and rapid methods of determining hydrogen ion concentrations, Am. Jour. Physiol., 1915, vol. 38, pp. 180–185.

[7] McClendon, J. F., A direct reading potentiometer for measuring hydrogen ion concentrations, Am. Jour. Physiol., 1915, vol. 38, pp. 186–190.

[8] McClendon, J. F., and Magoon, C. A., An improved Hasselbalch hydrogen electrode and a combined tonometer and hydrogen electrode, together with rapid methods of determining the buffer value of blood, Jour. Biol. Chem., 1916, vol. 25, pp. 669–681.

[9] Walpole, G. S., Gas-electrode for general use, Biochem. Jour., 1913, vol. 7, pp. 410–428.

[10] Walpole, G. S., An improved hydrogen electrode, Biochem. Jour., 1914, vol. 8, pp. 131–133.

[11] Wilke, E., Ueber eine neue Wasserstoffelektrode und ihre Verwendbarkeit, Zeitschr. f. Electrochem., 1913, vol. 19, pp. 857–858.

B. GENERAL METHODS.

[12] Baragiola, W. I., Concentration of hydrogen ions, Schweiz. Apoth. Ztg., 1914, vol. 52, pp. 641–643, quoted from Chemical Abstr., 1915, vol. 9, p. 349.

[13] Bjerrum, N., Die Theorie der alkalimetrischen und azidimetrischen Titrierungen, Sammlung chem. u. chem.-tech. Vorträge, 1914, vol. 21, pp. 1–128.

[14] Böttger, W., Die Anwendung des Elektrometers als Indikator beim Titrieren von Säuren und Basen, Zeitschr. phys. Chem., 1897, vol. 24, p. 253.

[15] Crozier, W. J., Rogers, W. B., and Harrison, B. I., Methods employed for determining the hydrogen-ion concentration in body fluids, Surg., Gyn., and Obstet., 1915, vol. 21, pp. 722–727.

[16] Czepinski, V., Einige Messungen an Gasketten, Zeitschr. f. anorg. Chem., 1902, vol. 30, pp. 1–17.

[17] Denham, H. G., The electrometric determination of the hydrolysis of salts, Jour. Chem. Soc., London, 1908, vol. 93, pp. 41–63.

[18] Denham, H. G., Anomalous behavior of the hydrogen electrode in solutions of lead salts, and the existence of univalent lead ions in aqueous solutions, Jour. Chem. Soc., London, 1908, vol. 93, pp. 424–427.

[19] Desha, L. J., and Acree, S. F., On difficulties in the use of the hydrogen electrode in the measurement of the concentration of hydrogen ions in the presence of organic compounds, Am. Chem. Jour., 1911, vol. 46, pp. 638–648.

[20] Enklaar, J. E., Eine störende Wirkung der Gaselektrode bei der Bestimmung der Wasserstoffionenkonzentration durch electrische Messung., Chem. Zentralbl., 1910, vol. 14, p. 852.

[21] Fox, C. J. J., On the constancy of the hydrogen gas electrode, Chem. News, 1909, vol. 100, p. 161.

[22] Haas, A. R. C., A simple and rapid method of studying respiration by the detection of exceedingly minute quantities of carbon dioxide, Science, n.s., 1916, vol. 44, pp. 105–108.

[23] Hasselbalch, K. A., Elektrometrische Reaktionsbestimmung kohlensäurehaltiger Flüssigkeiten, Biochem. Zeitschr., 1910, vol. 30, pp. 317–331.

[24] Hasselbalch, K. A., Determination electrometrique de la réaction des liquides renfermant de l'acide carbonique, C.-R. Lab. Carlsberg, 1911, vol. 10, pp. 69–84.

[25] Hasselbalch, K. A., Methods for the electrometric determination of the concentration of hydrogen ions in biological fluids, Biochem. Bull., 1912–1913, vol. 2, pp. 367–372.

[26] Hasselbalch, K. A., Verbesserte Methodik bei der elektrometrischen Reaktionsbestimmung biologischer Flüssigkeiten, Biochem. Zeitschr., 1913, vol. 49, pp. 451–457.

[27] Hildebrand, J. H., Some applications of the hydrogen electrode in analysis, research and teaching, Jour. Am. Chem. Soc., 1913, vol. 35, pp. 847–871.

[28] Kistiakowsky, W., Zur Methodik der Messung von Elektrodenpotentialen, Zeitschr. f. Elektrochem., 1908, vol. 14, pp. 113–121.

[29] Lundén, H., Amphoteric electrolytes, Jour. Biol. Chem., 1908, vol. 4, pp. 267–288.

[30] McClendon, J. F., The standardization of a new colorimetric method for the determination of the hydrogen ion concentration, CO_2 tension, and CO_2 and O_2 content of sea water, of animal heat, and of CO_2 of the air, with a summary of similar data on biocarbonate solutions in general, Jour., Biol. Chem., 1917, vol. 30, pp. 265–288.

[31] McClendon, J. F., Shedlov, A., and Thomson, W., Tables for finding the alkaline reserve of blood serum, in health and in acidosis, from the total CO_2 or the alveolar CO_2 or the P_H at known CO_2 tension, Jour. Biol. Chem., 1917, vol. 31, pp. 519–525.

[32] Michaelis, L., Die Wasserstoffionenkonzentration. Ihre Bedeutung für die Biologie und die Methoden ihrer Messung., 210 pp., Berlin, J. Springer, 1914.

[33] Muller, P. T., et Allemandet, H., Sur une électrode à alcali, Jour. Chim. Phys., 1907, vol. 5, pp. 532–556.

[34] Noyes, A. A., Report of the Committee on standard methods for determining small hydrogen-ion concentrations, 8th Intern. Cong. Appl. Chem., 1912, vol. 25, pp. 95–96.

[35] Osterhout, W. J. V., and Haas, A. R. C., A simple method of measuring photosynthesis, Science, n.s., 1918, vol. 47, pp. 420–422.

[36] Ringer, W. E., The rapid measurement of the hydrogen ion concentration of liquids, Chem. Weekblad., vol. 8, pp. 293–295, quoted from Chemical Abstr., 1911, vol. 5, p. 3364.

[37] Schmidt, C. L. A., Table of H^+ and OH^- concentrations corresponding to electromotive forces determined in gas-chain measurements, Univ. Calif. Publ. Physiol., 1909, vol. 3, pp. 101–113.

[38] Sörensen, S. P. L., Ueber die Messung und Bedeutung der Wasserstoffionenkonzentration bei biologischen Prozessen, Ergeb. d. Physiol., 1912, vol. 12, pp. 393–532.

[39] Symes, W. L., Graphic approximation to the value of Sörensen's P_H in terms of its integral part, Jour. Physiol., 1916, vol. 50, Proc. Physiol. Soc., pp. xxx–xxxi.

[40] Wagner, R. J., Estimation of the hydrion concentration of very small quantities of liquids, Biochem. Zeitschr., 1916, vol. 74, pp. 239–232, quoted from Chem. Abst., 1916, vol. 10, p. 2753.

C. THEORETICAL.

[41] Abegg, R., and Bose, E., Ueber den Einfluss gleichioniger Zusätze auf die elektromotorische Kraft von Konzentrationsketten und auf die Diffusionsgeschwindigkeit; Neutralsalzwirkung, Zeitschr. f. phys. Chem., 1899, vol. 30, pp. 545–555.

[42] Auerbach, F., Die Potentiale der wichtigsten Bezugselektroden, Zeitschr. f. Elektrochem., 1912, vol. 18, pp. 13–18.

[43] Clarke, W. F., Myers., C. N., and Acree, S. F., A study of the hydrogen electrode, of the calomel electrode and of contact potential, Jour. Phys. Chem., 1916, vol. 20, pp. 243–265.

[44] Ellis, J. H., The free energy of hydrochloric acid in aqueous solution, Jour. Am. Chem. Soc., 1916, vol. 38, pp. 737–762.

[45] Eucken, A., Ueber den stationären Zustand zwischen polarisierten Wasserstoffelektroden, Zeitschr. f. physik. Chem., 1907, vol. 59, pp. 72–117.

[46] Freundlich, H., and Mäkelt, E., Ueber den absoluten Nullpunkt des Potentials, Zeitschr. f. Electrochem., 1909, vol. 15, pp. 161–165.

[47] Glaser, L., Studien über die elektrolytische Zersetzung wässriger Lösungen, Zeitschr. f. Electrochem., 1898, vol. 4, pp. 355–359, 373–379, 424–428.

[48] Goodwin, H. M., Studien zur voltaschen Kette, Zeitschr. f. physik. Chem., 1894, vol. 13, pp. 577–656.

[49] Hardman, R. T., and Lapworth, A., Electromotive forces in alcohol. II, The hydrogen electrode in alcohol and the influence of water on its electromotive force, Jour. Chem. Soc., London, 1911, vol. 99, pp. 2242–2253.

[50] Hardman, R. T., and Lapworth, A., Electromotive forces in alcohol. III, Further experiments with the hydrogen electrode in dry and moist hydrogen chloride, Jour. Chem. Soc., London, 1912, vol. 101, pp. 2249–2255.

[51] Harned, H. S., The hydrogen- and hydroxyl-ion activities of solutions of hydrochloric acid, sodium and potassium hydroxides in the presence of neutral salts, Jour. Am. Chem. Soc., 1915, vol. 37, pp. 2460–2482.

[52] Harned, H. S., The hydrogen and chlorine ion activities of solutions of potassium chloride in 0.1 molal hydrochloric acid, Jour. Am. Chem. Soc., 1916, vol. 38, pp. 1986–1995.

[53] Henderson, L. J., A diagrammatic representation of equilibria between acids and bases in solution, Jour. Am. Chem. Soc., 1908, vol. 30, pp. 954–960.

[54] Henderson, P., Zur Thermodynamik der Flüssigkeitsketten, Zeitschr. phys. Chem., 1907, vol. 59, pp. 118–127.

[55] Henderson, P., Zur Thermodynamik der Flüssigkeitsketten, Zeitschr. phys. Chem., 1908, vol. 63, pp. 325–345.

[56] Kanolt, C. W., Ionization of water at 0°, 18°, and 25° derived from conductivity measurements of the hydrolysis of the ammonium salt of dikektotetrahydrothiazole, Jour. Am. Chem. Soc., 1907, vol. 29, pp. 1402–1416.

[57] Le Blanc, M., Die elektromotorischen Kräfte der Polarization, Zeitschr. f. phys. Chem., 1891, vol. 8, pp. 299–330; 1893, vol. 12, pp. 333–358.

[58] Lewis, G. N., Die Bestimmung der Ionenhydratation durch Messung von elektromotorischen Kräften, Zeitschr. f. Electrochem., 1908, vol. 14, pp. 509–510.

[59] Lewis, G. N., The activity of the ions and the degree of dissociation of strong electrolytes, Jour. Am. Chem. Soc., 1912, vol. 34, pp. 1631–1644.

[60] Lewis, G. N., and Randall, M., The free energy of oxygen, hydrogen and the oxides of hydrogen, Jour. Am. Chem. Soc., 1914, vol. 36, pp. 1969–1993.

[61] Lewis, G. N., Brighton, T. B., and Sebastian, R. L., A study of hydrogen and calomel electrodes, Jour. Am. Chem. Soc., 1917, vol. 39, pp. 2245–2261.

[62] Lewis, W. K., Eine Methode zur Berechnung von Ionenkonzentration aus Potentialmessungen von Konzentrationsketten, Zeitschr. phys. Chem., 1908, vol. 63, pp. 171–176.

[63] Loomis, N. E., and Acree, S. F., A study of the hydrogen electrode, of the calomel electrode and of contact potential, Am. Chem. Jour., 1911, vol. 46, pp. 585–620.

[64] Loomis, N. E., and Acree, S. F., The application of the hydrogen electrode to the measurement of the hydrolysis of aniline hydrochloride, and the ionization of acetic acid in the presence of neutral salts, Am. Chem. Jour., 1911, vol. 46, pp. 621–637.

[65] Loomis, N. E., and Acree, S. F., The effect of pressure on the hydrogen electrode, Jour. Am. Chem. Soc., 1916, vol. 38, pp. 2391–2396.

[66] Lorenz, R., and Mohn, A., Der Neutralpunkt der Wasserstoffelektrode, Zeitschr. f. phys. Chem., 1907, vol. 60, pp. 422–430.

[67] Lorenz, R., and Böhi, A., Beiträge zur Theorie der elektrolytischen Ionen. Die elektrolytische Dissociation des Wassers, Zeitschr. f. phys. Chem., 1909, vol. 66, pp. 733–751.

[68] Lorenz, R., Zur Frage des Nullpunktes der elektrochemischen Potentiale, Zeitschr. f. Electrochem., 1909, vol. 15, pp. 62–64.

[69] Lovén, J. M., Zur Theorie der Flüssigkeitsketten, Zeitschr. f. phys. Chem., 1896, vol. 20, pp. 593–600.

[70] Löwenherz, R., Ueber den Einfluss des Zusatzes von Äthylalkohol auf die elektrolytische Dissociation des Wassers, Zeitschr. f. phys. Chem., 1896, vol. 20, pp. 283–302.

[71] Michaelis, L., and Rona, P., Die Dissoziationskonstanten einiger sehr schwacher Säuren, insbesondere der Kohlenhydrate, gemessen auf elektrometrischem Wege, Biochem. Zeitschr., 1913, vol. 49, pp. 232–248.

[72] Myers, C. N., and Acree, S. F., A study of the hydrogen electrode, of the calomel electrode, and of contact potential, Am. Chem. Jour., 1913, vol. 50, pp. 396–411.

[73] Nernst, W., Zur Kinetik der in Lösung befindlichen Kërper, Zeitschr. f. phys. Chem., 1888, vol. 2, pp. 613–637.

[74] Nernst, W., Die electromotorische Wirksamkeit der Ionen, Zeitschr. f. phys. Chem., 1889, vol. 4, pp. 129–181.

[75] Nernst, W., Zur Dissociation des Wassers, Zeitschr. f. phys. Chem., 1894, vol. 14, pp. 155–156.

[76] Nernst, W., Ueber die Zahlenwerte einiger wichtiger physikochemischer Konstanten, Zeitschr. f. Electrochem., 1904, vol. 10, pp. 629–630.

[77] Ostwald, W., Die Dissociation des Wassers, Zeitschr. f. phys. Chem., 1893, vol. 11, pp. 521–528.

[78] Ostwald, W., Ueber die absoluten Potentiale der Metalle nebst Bemerkungen über Normalelektroden, Zeitschr. f. phys. Chem., 1900, vol. 35, pp. 333–339.

[79] Poma, G., Neutralsalzwirkung und Zustand der Ionen in Lösung, Zeitschr. f. phys. Chem., 1914, vol. 88, pp. 671–685.

[80] Poma, G., and Patroni, A., Einfluss der Neutralsalze auf den Zustand der Ionen in Lösung, Zeitschr. f. phys. Chem., 1914, vol. 87, pp. 196–214.

[81] Rothmund, V., Die Potentialdifferenzen zwischen Metallen und Elektrolyten, Zeitschr. f. phys. Chem., 1894, vol. 15, pp. 1–32.

[82] Sackur, O., Ueber den Einfluss gleichioniger Zusätze auf die elektromotorische Kraft von Flüssigkeiten. Ein Beitrag zur Kenntnis des Verhaltens starker Elektrolyte, Zeitschr. f. phys. Chem., 1901, vol. 38, pp. 129–162.

[83] Smale, F. J., Studien über Gasketten, Zeitschr. f. physik. Chem., 1894, vol. 14, pp. 577–621.

[84] Tolman, R. C., and Greathouse, L. H., The concentration of hydrogen ion in sulfuric acid, Jour. Am. Chem. Soc., 1912, vol. 34, pp. 364–369.

[85] Wilsmore, N. T. M., Ueber Elektroden-potentiale, Zeitschr. f. phys. Chem., 1900, vol. 35, pp. 291–332.

[86] Wilsmore, N. T. M., and Ostwald, W., Ueber Elektrodenpotentiale und absolute Potentiale, Zeitschr. f. phys. Chem., 1901, vol. 36, pp. 91–98.

D. CONTACT POTENTIAL.

[87] Abegg, R., and Cumming, A. C., Zur Eliminierung der Flüssigkeitspotentiale, Zeitschr. f. Electrochem., 1907, vol. 13, pp. 17–18.

[88] Bjerrum, N., Ueber die Gültigkeit der Planckschen Formel für das Diffusionspotential, Zeitschr. f. Electrochem., 1911, vol. 17, pp. 58–61.

[89] Bjerrum, N., Ueber die Elimination des Flüssigkeitspotentials bei Messungen von Elektrodenpotentialen, Zeitschr. f. Electrochem., 1911, vol. 17, pp. 389–393.

[90] Bjerrum, N., Ueber die Elimination des Diffusionspotentials zwischen verdünnten wässerigen Lösungen durch Einschalten einer konzentrierten Chlorkaliumlösung, Zeitschr. f. phys. Chem., 1905, vol. 53, p. 428.

[91] Chanoz, A. M., Recherches expérimentales sur les contacts liquides, Ann. de l'Université de Lyon, 1906, no. 18, pp. 1–99.

[92] Cumming, A. C., The elimination of potential due to liquid contact, Trans. Faraday Soc., 1906, vol. 2, pp. 213–221.

[93] Cumming, A. C., A simple equation for the calculation of the diffusion potential, Trans. Faraday Soc., 1912, vol. 8, pp. 86–93.

[94] Cumming, A. C., and Gilchrist, E., The effect of variations in the nature of the liquid boundary on the electromotive force, Trans. Faraday Soc., 1913, vol. 9, pp. 174–185.

[95] Johnson, K. R., Zur Nernst-Planckschen Theorie über die Potentialdifferenz zwischen verdünnten Lösungen, Ann. d. Physik., 1904, (4), vol. 4, pp. 995–1003.

[96] Lewis, G. N., and Sargent, L. W., Potentials between liquids, Jour. Am. Chem. Soc., 1909, vol. 31, pp. 363–367.

[97] Negbaur, W., Experimentaluntersuchungen über Potentialdifferenzen an den Berührungsflächen sehr verdünnter Lösungen, Wiedemann's Ann. d. Phys. u. Chem., 1891, N.F., vol. 44, pp. 737–758.

[98] Planck, M., Ueber die Erregung von Electricität und Wärme in Electrolyten, Wiedemann's Ann. d. Phys. u. Chem., 1890, N.F., vol. 39, pp. 161–186.

[99] Planck, M., Ueber die Potentialdifferenz zwischen zwei verdünnten Lösungen binärer Electrolyte, Wiedemann's Ann. d. Phys. u. Chem., 1890, N.F., vol. 40, pp. 561–576.

[100] Tower, O. F., Ueber Potentialdifferenzen an den Berührungsflächen verdünnter Lösungen, Zeitschr. f. phys. Chem., 1896, vol. 20, pp. 198–206.

E. CALOMEL ELECTRODES

[101] Coggeshall, G. W., Ueber die Konstanz der Kalomelelektrode, Zeitschr. f. phys. Chem., 1895, vol. 17, pp. 62–86.

[102] Desha, L. J., An apparatus for the purification of mercury, Am. Chem. Jour., 1909, vol. 41, pp. 152–155.

[103] Hildebrand, J. H., Purification of mercury, Jour. Am. Chem. Soc., 1909, vol. 31, pp. 933–935.

[104] Hulett, G. A., and Minchin, H. D., The distillation of amalgams and the purification of mercury, Physical Rev., 1905, vol. 21, pp. 388–398.

[105] Lipscomb, G. F., and Hulett, G. A., A calomel standard cell, Jour. Am. Chem. Soc., 1916, vol. 38, p. 20.

[106] Loomis, N. E., Notes upon the potentials of calomel and hydrogen electrodes, Jour. Phys. Chem., 1915, vol. 19, pp. 660–664.

[107] Palmaer, W., Ueber das absolute Potential der Kalomelelektrode, Zeitschr. f. phys. Chem., 1907, vol. 59, pp. 129–191.

[108] Palmaer, W., Ueber das absolute Potential der Kalomelelektrode, Zeitschr. f. Electrochem., 1903, vol. 9, pp. 754–757.

[109] Richards, T. W., Ueber den Temperaturkoëffizienten des Potentials der Kalomelelektrode mit verschiedenen gelösten Elektrolyten, Zeitschr. f. phys. Chem., 1897, vol. 24, pp. 39–54.

[110] Sauer, L., Bezugselektroden, Zeitschr. f. phys. Chem., 1904, vol. 47, pp. 146–184.

F. BUFFER MIXTURES.

[111] Clark, W. M., and Lubs, L. A., Hydrogen electrode potentials of phthalate, phosphate, and borate buffer mixtures, Jour. Biol. Chem., 1916, vol. 25, pp. 479–510.

[112] Koppel, M., and Spiro, K., Ueber die Wirkung von Moderatoren (Puffern) bei der Verschiebung des Säure-Basengleichgewichtes in biologischen Flüssigkeiten, Biochem. Zeitschr., 1914, vol. 65, pp. 409–439.

[113] Palitzsch, S., Ueber die Anwerdung von Borax- und Borsäurelösungen bei der colorimetrischen Messung der Wasserstoffionenkonzentration des Meerwassers, Biochem. Zeitschr., 1915, vol. 70, pp. 333–343.

[114] Palitzsch, S., Sur l'emploi de solutions de borax et d'acide borique dans la determination colorimétrique de la concentration en ions hydrogène de l'eau de mer, C.-R. Lab. Carlsberg, 1916, vol. 11, pp. 199–211.

[115] Prideaux, E. B. R., The sodium phosphate standards of acidity, Biochem. Jour., 1911, vol. 6, pp. 122–126.

[116] Prideaux, E. B. R., On the use of partly neutralized mixtures of acids as hydrion regulators, Proc. Royal Soc., London (A), 1916, vol. 92, pp. 463–468.

[117] Ringer, W. E., The concentration of hydrogen ions in dilute solutions of phosphoric acid, mono- and disodium phosphate, Chem. Weekblad, vol. 6, pp. 446–452, quoted from Chemical Abst., 1910, vol. 4, pp. 5–6.

[118] Schmidt, C. L. A., and Finger, C. P., Potential of a hydrogen electrode in acid and alkaline solutions, Jour. Phys. Chem., 1908, vol. 12, pp. 406–416.

[119] Walpole, G. S., The effect of dilution on the hydrogen potentials of acetic acid and "standard acetate" solutions, Jour. Chem. Soc., London, 1914, vol. 105, pp. 2521–2529.

[120] Walpole, G. S., Hydrogen potentials of mixtures of acetic acid and sodium acetate, Jour. Chem. Soc., London, 1914, vol. 105, pp. 2501–2520.

[121] Walpole, G. S., Notes on regular mixtures, recent indicators, etc. II, Biochem. Jour., 1914, vol. 8, pp. 628–640.

G. MISCELLANEOUS ELECTRODES.

[122] Bancroft, W., Ueber Oxydationsketten, Zeitschr. f. phys. Chem., 1892, vol. 10, pp. 387–409.

[123] Bose, E., Untersuchungen über die elektromotorische Wirksamkeit der elementären Gase, Zeitschr. f. phys. Chem., 1900, vol. 34, pp. 700–760.

[124] Bose, E., Untersuchungen über die elektromotorische Wirksamkeit der elementären Gase, Zeitschr. f. phys. Chem., 1901, vol. 38, pp. 1–26.

[125] Brislee, F. J., The potential of the hydrogen-oxygen cell, Trans. Faraday Soc., 1905, vol. 1, pp. 65–74.

[126] Brönsted, J. N., Die electromotorische Kraft der Knallgaskette, Zeitschr. f. phys. Chem., 1908, vol. 65, pp. 84–92.

[127] Hoeper, V., Ueber die elektromotorische Wirksamkeit des Kohlenoxyd-gases, Zeitschr. f. anorg. Chem., 1899, vol. 20, pp. 419–451.

[128] Lapworth, A., and Partington, J. R., Electromotive forces in alcohol. I, Concentration cells with electrodes reversible to chlorine ions, Jour. Chem. Soc., London, 1911, vol. 99, pp. 1417–1427.

[129] Laurie, A. P., Die elektromotorische Kraft von Iodkonzentrationsketten in Wasser und Alkohol, Zeitschr. phys. Chem., 1908, vol. 64, pp. 615–628.

[130] Lewis, G. N., and Sargent, L. W., The potential of the ferro-ferricyanide electrode, Jour. Am. Chem. Soc., 1909, vol. 31, pp. 355–363.

[131] Lewis, G. N., and Rupert, F. F., The potential of the chlorine electrode, Jour. Am. Chem. Soc., 1911, vol. 33, pp. 299–307.

[132] Lorenz, R., Die Oxydtheorie der Sauerstoffelektrode, Zeitschr. f. Electrochem., 1908, vol. 14, pp. 781–783.

[133] Luther, R., and Michie, A. C., Das elektromotorische Verhalten von Uranyl-Uranogemengen, Zeitschr. f. Electrochem., 1908, vol. 14, pp. 826–829.

[134] Maitland, W., Ueber das Iod-Potential und das Ferri-Ferro-Potential, Zeitschr. f. Electrochem., 1906, vol. 12, pp. 263–268.

[135] Müller, E., Die elektromotorische Kraft der Chlorknallgaskette, Zeitschr. f. phys. Chem., 1902, vol. 40, pp. 158–168.

[136] Naumann, R., Die elektromotorische Kraft der Cyanwasserstoffkette, Zeitschr. f. Elektrochem., 1910, vol. 16, pp. 191–199.

[137] Nernst, W., and Sand, J., Zur Kenntnis der unterchlorigen Säure. Elektromotorisches Verhalten, Zeitschr. f. phys. Chem., 1904, vol. 48, pp. 601–609.

[138] Nernst, W., and Wartenberg, H. v., Die Dissociation von Wasserdampf, Zeitschr. phys. Chem., 1906, vol. 56, pp. 513–547.

[139] Neumann, B., Ueber das Potential des Wasserstoffs und einiger Metalle, Zeitschr. f. phys. Chem., 1894, vol. 14, pp. 193–230.

[140] Peters, R., Ueber Oxydations- und Reduktions-ketten und den Einfluss komplexer Ionen auf ihre elektromotorische Kraft, Zeitschr. f. phys. Chem., 1898, vol. 26, pp. 193–236.

[141] Preuner, G., Ueber die Dissociationskonstante des Wassers und die elektromotorische Kraft der Knallgaskette, Zeitschr. f. phys. Chem., 1902, vol. 42, pp. 50–58.

[142] Schoch, E. P., A study of reversible oxidation and reduction reactions in solutions, Jour. Am. Chem. Soc., 1904, vol. 26, pp. 1422–1433.

[143] Schoch, E. P., The potential of the oxygen electrode: a report of progress, Jour. Phys. Chem., 1910, vol. 14, pp. 665–677.

II. Biological.

A. Blood.

[144] Abel, E., and Fürth, O. v., Zur physikalischen Chemie des Oxyhämoglobins. Das Alkalibindungsvermögen des Blutfarbstoffes, Zeitschr. f. Electrochem., 1906, vol. 12, pp. 349–359.

[145] Aggazzotti, A., La reazione de sangue nell' aria rarefatta determinata coi metodi titolimetrici ed elettrometrici, Atti d. Reale Accad. d. Lincei, Rome, 1906, vol. 15, pp. 474–483.

[146] Benedict, H., Der Hydroxylionengehalt des Diabetikerblutes, Arch. f. d. ges. Physiol., 1906, vol. 115, pp. 106–117.

[147] Bugarszky, S., and Tangl, F., Physikalisch-chemische Untersuchungen über die molecularen Concentrationsverhältnisse des Blutserums, Arch. f. d. ges. Physiol., 1898, vol. 72, pp. 531–565.

[148] Corral, J. M., Ueber die elektrometrische Bestimmung der wahren Reaktion des Blutes, Biochem. Zeitschr., 1915, vol. 72, pp. 1–25.

[149] Cullen, G. E., The electrometric titration of plasma as a measure of its alkaline reserve, Jour. Biol. Chem., 1917, vol. 30, pp. 369–388.

[150] Farkas, G., Ueber die Concentration der Hydroxylionen im Blutserum, Arch. f. d. ges. Physiol., 1903, vol. 98, pp. 551–576.

[151] Farkas, G., and Scipiades, E.. Ueber die molekularen Concentrations-verhältnisse des Blutserums der Schwangeren, Kreissenden und Wöchnerinnen und des Fruchtwassers, Arch. f. d. ges. Physiol., 1903, vol. 98, pp. 577–587.

[152] Fraenckel, P., Eine neue Methode zur Bestimmung der Reaction des Blutes, Arch. f. d. ges. Physiol., 1903, vol. 96, pp. 601–623.

[153] Friedenthal, H., Ueber die Reaktion des Blutserums der Wirbeltiere und die Reaktion der lebendigen Substantz im allgemeinen, Zeitschr. f. allgemeine Physiol., 1902, vol. 1, pp. 56–66.

[154] Friedenthal, H., Reactionsbestimmungen im natürlichen Serum und über Herstellung einer zum Ersatz des natürlichen Serums geeigneten Salzlösung, Arch. Anat. u. Physiol. (Physiol. Abt.), 1903, pp. 550–554.

[155] Friedenthal, H., Ueber die Reaktion des Blutserums der Wirbeltiere und die Reaktion der lebendigen Substanz im allgemeinen, Zeitschr., f. allgemeine Physiol., 1904, vol. 4, pp. 44–61.

[156] Hasselbalch, K. A., The reduced and the regulated "hydrogen figure" of the blood, Biochem. Zeitschr., 1916, vol. 74, pp. 56–62, quoted from Physiol Abst., 1916, vol. 1, p. 252.

[157] Hasselbalch, K. A., The calculation of the hydrogen number of the blood from the free and bound carbon dioxide of the same, and the binding of oxygen by the blood as a function of the hydrogen number, Biochem. Zeitschr., 1916, vol. 78, pp. 112–144, quoted from Chem. Abst., 1917, vol. 11, pp. 1656–1657.

[158] Hasselbalch, K. A., The true nature of the "acidotic condition" of infants, Biochem. Zeitschr., 1917, vol. 80, pp. 251–258, quoted from Chem. Abst., 1917, vol. 11, p. 2693.

[159] Hasselbalch, K. A., and Gammeltoft, S. A., Die Neutralitätsregulation des graviden Organismus, Biochem. Zeitschr., 1914, vol. 68, pp. 206–264.

[160] Hasselbalch, K. A., and Lundsgaard, C., Blutreaktion und Lungenventilation, Skand. Arch. f. Physiol., 1912, vol. 27, pp. 13–31.

[161] Hasselbalch, K. A., and Lundsgaard, C., Elektrometrische Reaktionsbestimmung des Blutes bei Körpertemperatur, Biochem. Zeitschr., 1912, vol. 38, pp. 77–91.

[162] Henderson, L. J., On the neutrality equilibrium in blood and protoplasm, Jour. Biol. Chem., 1909, vol. 7, pp. 29–35.

[163] Höber, R., Ueber die Hydroxylionen des Blutes, Arch. f. d. ges. Physiol., 1900, vol. 81, pp. 522–539.

[164] Höber, R., Ueber die Hydroxylionen des Blutes, Arch. f. d. ges. Physiol., 1903, vol. 99, pp. 572–593.

[165] Höber, R., Die Gaskettenmethode zur Bestimmung der Blutreaktion, Deut. med. Woch., 1917, vol. 43, pp. 551–552.

[166] Homer, A., A note on the use of indicators for the colorimetric determination of the hydrogen ion concentration of sera, Biochem. Jour., 1917, vol. 11, pp. 283–291.

[167] Konikoff, A. P., Ueber die Bestimmung der wahren Blutreaktion mittels der elektrischen Methode, Biochem. Zeitschr., 1913, vol. 51, pp. 200–210.

[168] Kreibich, C., Ueber die Hydroxylionenkonzentration des pathologischen Blutes, Wien. klin. Woch., 1910, vol. 23, pp. 355–358; 1911, vol. 24, pp. 1419–1420.

[169] Levy, R. L., Rowntree, L. G., and Marriott, W. M., A simple method for determining variations in the hydrogen-ion concentration of the blood, Arch. Int. Med., 1915, vol. 16, pp. 389–405.

[170] Lundsgaard, C., Die Reaktion des Blutes, Biochem. Zeitschr., 1912, vol. 41, pp. 247–267.

[171] Masel, J., Zur Frage der Säurevergiftung beim Coma diabeticum, Zeitschr. f. klin. Med., 1914, vol. 79, pp. 1–12.

[172] McClendon, J. F., Improved gas chain methods of determining hydrogen ion concentration in blood, Jour. Biol. Chem., 1916, vol. 24, pp. 519–526.

[173] McClendon, J. F., A new hydrogen electrode for the electrometric titration of the alkaline reserve of blood plasma and other frothing fluids, Jour. Biol. Chem., 1918, vol. 33, pp. 19–29.

[174] Menten, M. L., and Crile, G. W., Studies on the hydrogen-ion concentration in blood under various abnormal conditions, Am. Jour. Physiol., 1915, vol. 38, pp. 225–232.

[175] Michaelis, L., Die Bedeutung des Wasserstoffionenkonzentration des Blutes und der Gewebe, Deut. med. Woch., 1914, vol. 40, pp. 1170–1171.

[176] Michaelis, L., and Davidoff, W., Methodisches und Sachliches zur elektrometrischen Bestimmung der Blutalkalescenz, Biochem. Zeitschr., 1912, vol. 46, pp. 131–150.

[177] Michaelis, L., and Rona, P., Electrochemische Alkalinitätsmessungen an Blut und Serum, Biochem. Zeitschr., 1909, vol. 18, pp. 317–339.

[178] Milroy, T. H., Changes in the hydrogen ion concentration of the blood produced by pulmonary ventilation, Quar. Jour. Exper. Physiol., 1914, vol. 8, pp. 141–153.

[179] Palmer, W. W., and Henderson, L. J., Clinical studies on acid base equilibrium and the nature of acidosis, Arch. Int. Med., 1913, vol. 12, pp. 153–170.

[180] Peters, R. A., A combined tonometer and electrode cell for measuring the H-ion concentration of reduced blood at a given tension of CO_2, Jour. Physiol., Proc. Physiol. Soc., 1914, vol. 48, pp. vii, viii.

[181] Pfaundler, M., Ueber die actuelle Reaction des kindlichen Blutes, Arch. v. Kinderheilk., 1905, vol. 41, pp. 161–184.

[182] Porges, O., The "reduced" and "regulated" hydrogen number of the blood, Biochem. Zeit., 1916, vol. 77, pp. 241–248, quoted from Physiol. Abst., 1917, vol. 2, pp. 32–33.

[183] Quagliariello, G., Ueber die Hydroxylionenkonzentration des Blutes bei der Temperaturerhöhung nach dem Wärmestich, Biochem. Zeitschr., 1912, vol. 44, pp. 162–164.

[184] Robertson, T. B., On the nature of the chemical mechanism which maintains the neutrality of the tissues and tissue-fluids, Jour. Biol. Chem., 1909, vol. 6, pp. 313–320.

[185] Robertson, T. B., Concerning the relative magnitude of the parts played by the proteins and by the bicarbonates in the maintenance of the neutrality of the blood, Jour. Biol. Chem., 1910, vol. 7, pp. 351–357.

[186] Rolly, F., Ueber die Reaktion des Blutserums bei normalen und pathologischen Zuständen, Münch. med. Woch., 1912, vol. 59, pp. 1201–1205, 1274–1277.

[187] Rona, P., and György, P., Beitrag zur Frage der Ionenverteilung im Blutserum, Biochem. Zeitschr., 1913, vol. 56, pp. 416–438.

[188] Rona, P., and Takahashi, D., Beitrag zur Frage nach dem Verhalten des Calciums im Serum, Biochem. Zeitschr., 1913, vol. 49, pp. 370–380.

[189] Rona, P., and Yippö, A., Ueber den Einfluss der Wasserstoffionenkonzentration auf die Saurestoffdissoziationscurve des Hämoglobins, Biochem. Zeitschr., 1916, vol. 76, pp. 187–217, quoted from Physiol. Abst., 1917, vol. 2, p. 32.

[190] Scott, R. W., The effect of the accumulation of carbon dioxide on the tidal air and on the H-ion concentration of the arterial blood in the decerebrate cat, Am. Jour. Physiol., 1917, vol. 44, pp. 196–211.

[191] Stillman, E., Van Slyke, D. D., Cullen, G. E., and Fitz, R., The blood, urine, and alveolar air in diabetic acidosis, Jour. Biol. Chem., 1917, vol. 30, pp. 405–456.

[192] Szili, A., Untersuchungen über den Hydroxylionengehalt des placentaren (fötalen) Blutes, Arch. f. d. ges. Physiol., 1906, vol. 115, pp. 72–81.

[193] Winternitz, H., Beiträge zur Alkalimetrie des Blutes, Zeitschr. f. physiol. Chem., 1891, vol. 15, pp. 505–512.

B. URINE.

[194] Blatherwick, N. R., The specific rôle of foods in relation to the composition of the urine, Arch. Int. Med., 1914, vol. 14, pp. 409–450.

[195] Bugarszky, S., Ueber die molecularen Concentrationsverhältnisse des normalen menschlichen Harns, Arch. f. d. ges. Physiol., 1897, vol. 68, pp. 389–407.

[196] Foà, C., Le réaction de l'urine et du suc pancréatique étudieé par la méthode electrométrique, C.-R. Soc. Biol., 1905, vol. 58, pp. 867–869.

[197] Henderson, L. J., Messungen der normalen Harnacidität, Biochem. Zeitschr., 1910, vol. 24, pp. 40–44.

[198] Henderson, L. J., and Palmer, W. W., On the intensity of urinary acidity in normal and pathological conditions, Jour. Biol. Chem., 1913, vol. 13, pp. 393–405.

[199] Henderson, L. J., and Palmer, W. W., On the extremes of variation of the concentration of ionized hydrogen in human urine, Jour. Biol. Chem., 1913, vol. 14, pp. 81–85.

[200] Henderson, L. J., and Spiro, K., Ueber Basen- und Säuregleichgewicht im Harn, Biochem. Zeitschr., 1908, vol. 15, pp. 105–113.

[201] Höber, R., Die Acidität des Harns vom Standpunkt der Ionenlehre, Beitr. z. chem. Physiol. u. Path., 1903, vol. 3, pp. 525–542.

[202] Howe, P. E., and Hawk, P. B., Variations in the hydrogen ion concentration of the urine of man accompanying fasting and the low- and high-protein regeneration periods, Jour. Biol. Chem., Proc., 1914, vol. 17, p. xlviii.

[203] Newburgh, L. H., Palmer, W. W., and Henderson, L. J., A study of hydrogen ion concentration of the urine in heart disease, Arch. Int. Med., 1913, vol. 12, pp. 146–152.

[204] Rhorer, L. v., Die Bestimmung der Harnacidität auf elektrometrischen Wege, Arch. f. d. ges. Physiol., 1901, vol. 86, pp. 586–602.

[205] Ringer, W. E., Zur Acidität des Harns, Zeitschr. f. physiol. Chem., 1909, vol. 60, pp. 341–363.

[206] Ringer, W. E., Ueber die Bedingungen der Ausscheidung von Harnsäure und harnsäuren Salzen aus ihren Lösungen, Zeitschr. f. physiol. Chem., 1910, vol. 67, pp. 332–403.

[207] Skramlik, E. v., Ueber Harnacidität, Zeitschr. f. physiol. Chem., 1911, vol. 71, pp. 290–310.

C. MILK.

[208] Allemann, O., Die Bedeutung der Wasserstoffionen für die Milchgerinnung, Biochem. Zeitschr., 1912, vol. 45, pp. 346–358.

[209] Clark, W. M., The reaction of cow's milk modified for infant feeding, Jour. Med. Res., 1915, vol. 31, pp. 431–453.

[210] Davidsohn, H., Ueber die Reaktion der Frauenmilch, Zeitschr. f. Kinderheilk., 1913, vol. 9, pp. 11–18.

[211] Foà, C., La réaction du lait et de l'humeur aqueuse étudiée par la méthode electrométrique, C.–R. Soc. Biol., 1905, vol. 59, pp. 51–53.

[212] Michaelis, L., and Mendelssohn, A., Die Wirkungsbedingungen des Labferments, Biochem. Zeitschr., 1913, vol. 58, pp. 315–328.

D. MISCELLANEOUS BODY FLUIDS AND TISSUES.

[213] Allaria, G. B., Untersuchungen über Wasserstoff-Ionen-Konzentration im Säuglingsmagen, Jahr. f. Kinderheilk., Ergänzungsband 1908, vol. 67, pp. 123–142.

[214] Auerbach, F., and Pick, H., Die Alkalität von Pankreassaft und Darmsaft lebender Hunde, Arbeit. a. d. kais. Gesundheitsamte, 1913, vol. 43, pp. 155–186.

[215] Bisgaard, A., Untersuchungen über die Eiweiss- und Stickstoffverhältnisse der Cerebrospinalflüssigkeit sowie über die Wasserstoffionenkonzentration derselben, Biochem. Zeitschr., 1913, vol. 58, pp. 1–64.

[216] Christiansen, J., Die Sauerstoffionen-Konzentration im Mageninhalt, Deut. Arch. f. klin. Med., 1911, vol. 102, pp. 103–116.

[217] Christiansen, J., Bestimmung freier Salzsäure im Mageninhalt, Biochem. Zeitschr., 1912, vol. 46, pp. 24–49.

[218] Davidsohn, H., Beitrag zum Chemismus des Säuglingsmagens, Zeitschr. f. Kinderheilk., 1911, vol. 2, pp. 420–428.

[219] Felton, L. D., Hussey, R. G., and Bayne-Jones, S., The reaction of the cerebrospinal fluid, Arch Int. Med., 1917, vol. 19, pp. 1085–1096.

[220] Foä, C., La réaction des liquides de l'organisme étudiée par la méthode electrométrique, C.–R. Soc. biol., 1905, vol. 58, pp. 865–866.

[221] Foä, C., La réaction de quelques liquides de l'organisme étudiée par la méthode electrométrique, C.–R. Soc. Biol., 1905, vol. 58, pp. 1000–1002.

[222] Foà, C., La réaction du suc gastrique; étudiée par la méthode electrométrique, C.–R. Soc. Biol., 1905, vol. 59, pp. 2–4.

[223] Foà, C., La reazione dei liquidi dell' organismo determinata col metodo elettrometrico (pile di concentrazione), Arch. di Fisiologica, 1906, vol. 3, pp. 369–415.

[224] Fraenckel, P., Die Wasserstoffionenconcentration des reinen Magensaftes und ihre Beziehung zur elektrischen Leitfähigkeit und zur titrimetrischen Acidität, Zeitschr. exper. Path. u. Therap., 1905, vol. 1, pp. 431–438.

[225] Fraenckel, P., Ueber den Einfluss der Erdalkalien auf die Reaction thierischer Säfte, Zeitschr. exper. Path. u. Therap., 1905, vol. 1, pp. 439–445.

[226] Galeotti, G., Sui fenomeni elettrici del cuore. Variazioni della concentrazione degli H-ioni nel mioplasma durante la contrazione o nella morte del cuore, Arch. di Fisiologica, 1904, vol. 1, pp. 512–530.

[227] Henderson, L. J., The theory of neutrality regulation in the animal organism, Am. Jour. Physiol., 1908, vol. 21, pp. 427–448.

[228] Henderson, L. J., Das Gleichgewicht zwischen Basen und Säuren im tierischen Organismus, Ergeb. d. Physiol., 1909, vol. 8, pp. 254–325.

[229] Henderson, L. J., A critical study of the process of acid excretion, Jour. Biol. Chem., 1911, vol. 9, pp. 403–424.

[230] Hess, R., Die Acidität des Säuglingsmagens, Zeitschr. f. Kinderheilk., 1915, vol. 12, pp. 409–439.

[231] Höber, R., Die Beteiligung von Wasserstoff- und Hydroxylionen bei physiologischen Vorgängen, Zeitschr. Electrochem., 1910, vol. 16, pp. 681–686.

[232] Howe, P. E., and Hawk, P. B., Hydrogen ion concentration of feces, Jour. Biol. Chem., 1912, vol. 11, pp. 129–140.

[233] Hurwitz, S. H., and Tranter, C. L., On the reaction of the cerebrospinal fluid, Arch. Int. Med., 1916, vol. 17, pp. 828–839.

[234] Löb, W., and Higuchi, S., Die Wasserstoff- und Hydroxylionenkonzentrationen des Placentar- und Retroplacentarserums, Biochem. Zeitschr., 1910, vol. 24, pp. 92–107.

[235] Long, J. H., and Fenger, F., On the reaction of the pancreas, Jour. Am. Chem. Soc., 1915, vol. 37, pp. 2213–2219.

[236] Long, J. H., and Fenger, F., On the reaction of the pancreas and other organs, Jour. Am. Chem. Soc., 1916, vol. 38, pp. 1115–1128.

[237] McClendon, J. F., Acidity curves in the stomachs and duodenum of adults and infants, plotted with the aid of improved methods of measuring hydrogen ion concentration, Am. Jour. Physiol., 1915, vol. 38, pp. 191–199.

[238] McClendon, J. F., Shedlov, A., and Thomson, W., The hydrogen ion concentration of the ileum content, Jour. Biol. Chem., 1917, vol. 31, pp. 269–270.

[239] McClendon, J. F., Shedlov., A., and Karpman, B., The hydrogen ion concentration of the contents of the small intestine, Jour. Biol. Chem., 1918, vol. 34, pp. 1–3.

[240] Menten, M. L., Acidity of undiluted normal gastric juice from a case of human gastric fistula, Jour. Biol. Chem., 1915, vol. 22, pp. 341–343.

[241] Michaelis, L., and Davidsohn, H., Die Bedeutung und die Messung der Magensaftacidität, Zeitschr. exper. Path. u. Therap., 1910, vol. 8, pp. 398–413.

[242] Michaelis, L., and Kramsztyk, A., Die Wasserstoffionenkonzentration der Gewebssäfte, Biochem. Zeitschr., 1914, vol. 62, pp. 180–186.

[243] Okada, S., On the reaction of bile, Jour. Physiol., 1915, vol. 50, pp. 114–118.

[244] Quagliariello, G., Sulla reazione chimica della bile, Atti d. Reale Accad. d. Lincei, Rome, 1911, vol. 20, pp. 302–305.

[245] Rosemann, R., Physiologie der Verdauung. VII, H-ionen Konzentration des Magensaftes., Arch. f. d. ges. Physiol., 1917, vol. 169, pp. 188–200, quoted from Physiol. Abst., 1918, vol. 2, p. 604.

[246] Salge, B., Salzsäure im Säuglingsmagen, Zeitschr. f. Kinderheilk., 1912, vol. 4, pp. 171–173.

[247] Tangl, F., Untersuchungen über die Hydrogenionenkonzentration im Inhalte des nüchternen menschlichen Magens, Arch. f. d. ges. Physiol., 1906, vol. 115, pp. 64–71.

E. PROTEINS.

[248] Bugarszky, S., and Liebermann, L., Ueber das Bindungsvermögen eiweissartiger Körper für Salzsäure, Natriumhydroxyd und Kochsalz, Arch. f. d. ges. Physiol., 1898, vol. 72, pp. 51–74.

[249] Chick, H., and Martin, C. J., On the ''heat coagulation'' of proteins, Jour. Physiol., 1910, vol. 40, pp. 404–430; 1911, vol. 43, pp. 1–27.

[250] Chick, H., and Martin, C. J., Die Hitzekoagulation der Eiweisskörper, Kolloid. chem. Beihefte, 1913, vol. 5, pp. 49–140.

[251] Henderson, L. J., Palmer, W. W., and Newburgh, L. H., The swelling of colloids and hydrogen ion concentration, Jour. Pharm. and Exper. Therap., 1914, vol. 5, pp. 449–467.

[252] Manabe, K., and Matula, J., Elektrochemische Untersuchungen am Säureeiweiss, Biochem. Zeitschr., 1913, vol. 52, pp. 369–408.

[253] Michaelis, L., Die elektrische Ladung des Serumalbumins und der Fermente, Biochem. Zeitschr., 1909, vol. 19, pp. 181–185.

[254] Michaelis, L., Ueber die Dissoziation der amphoteren Elektrolyte, Biochem. Zeitschr., 1911, vol. 33, pp. 182–189.

[255] Michaelis, L., Zur Theorie des isoelektrischen Punktes. III, Das Wesen der Eiweissartigen kolloidalen Lösungen, Biochem. Zeitschr. 1912, vol. 47, pp. 250–259.

[256] Michaelis, L., and Davidsohn, H., Der isoelektrische Punkt des gemeinen und des denaturierten Serumalbumins, Biochem. Zeitschr., 1911, vol. 33, pp. 456–473.

[257] Michaelis, L., and Davidsohn, H., Ueber das Flockungsoptimum von Kolloidgemischen, Biochem. Zeitschr., 1912, vol. 39, pp. 496–506.

[258] Michaelis, L., and Davidsohn, H., Ueber die Kataphorese des Oxyhämoglobins, Biochem. Zeitschr., 1912, vol. 41, pp. 102–110.

[259] Michaelis, L., and Davidsohn, H., Die Abhängigkeit spezifischer Fällungsreaktionen von der Wasserstoffionenkonzentration, Biochem. Zeitschr., 1912, vol. 47, pp. 59–72.

[260] Michaelis, L., and Davidsohn, H., Zur Theorie des isoelektrischen Punktes, Biochem. Zeitschr., 1910, vol. 30, pp. 143–150.

[261] Michaelis, L., and Davidsohn, H., Weiterer Beitrag zur Frage nach der Wirkung der Wasserstoffionenkonzentration auf Kolloidgemische, Biochem. Zeitschr., 1913, vol. 54, pp. 323–329.

[262] Michaelis, L., and Grineff, W., Der isoelektrische Punkt der Gelatine, Biochem. Zeitschr., 1912, vol. 41, pp. 373–374.

[263] Michaelis, L., and Mostynski, B., Die isoelektrische Konstante und die relative Aciditätskonstante des Serumalbumins, Biochem. Zeitschr., 1910, vol. 24, pp. 79–91.

[264] Michaelis, L., and Mostynski, B., Die innere Reibung von Albuminlösungen, Biochem. Zeitschr., 1910, vol. 25, pp. 401–410.

[265] Michaelis, L., and Pechstein, H., Der isoelektrische Punkt des Caseins, Biochem. Zeitschr., 1912, vol. 47, pp. 260–268.

[266] Michaelis, L., and Rona, P., Die Koagulation des denaturierten Albumins als Funktion der Wasserstoffionenkonzentration und der Salze, Biochem Zeitschr., 1910, vol. 27, pp. 38–52.

[267] Michaelis, L., and Rona, P., Ueber die Fallung des Globuline im isoelektrischen Punkt, Biochem. Zeitschr., 1910, vol. 28, pp. 193–199.

[268] Quagliariello, G., Die Änderung der Wasserstoffionenkonzentration während der Hitzkoagulation der Proteine, Biochem. Zeitschr., 1912, vol. 44, pp. 157–161.

[269] Roberson, T. B., On the dissociation of serum globulin at varying hydrogen ion concentrations, Jour. Phys. Chem., 1907, vol. 11, pp. 437–460.

[270] Robertson, T. B., The dissociation of potassium caseinate in solutions of varying alkalinity, Jour. Phys. Chem., 1910, vol. 14, pp. 528–568.

[271] Robertson, T. B., The dissociation of the salts of ovomucoid in solutions of varying alkalinity and acidity, Jour. Phys. Chem., 1910, vol. 14, pp. 709–718.

[272] Schmidt, C. L. A., Changes in the H^+ and OH^- concentration which take place in the formation of certain protein compounds, Jour. Biol. Chem., 1916, vol. 25, pp. 63–79.

[273] Schmidt, C. L. A., Studies on the formation and antigenic properties of certain compound proteins, Univ. Calif. Publ. Path., 1916, vol. 2, pp. 157–204.

[274] Spiro, K., Die Fällung von Kolloiden, Biochem. Zeitschr., 1913, vol. 56, pp. 11–16.

[275] Sörensen, S. P. L., and Jürgensen, E., La concentration en ions hydrogène du milieu subit-elle des modifications par suite de la coagulation? C.-R. Lab., Carlsberg, 1911, vol. 10, pp. 1–51.

[276] Sörensen, S. P. L., and Jürgensen, E., Wird die Wasserstoffionenkonzentration der Lösung, durch die Koagulation geändert, Biochem. Zeitschr., 1911, vol. 31, pp. 397–442.

[277] Ylppö, A., Der isoelektrische Punkt des Menschen- Kuh- Ziegen- Hunde- und Meerschweinschenmilchcaseins, Zeitschr. f. Kinderheilk., 1913, vol. 8, pp. 224–234.

F. ENZYMES.

[278] Adler, L., The influence of the hydrion concentration on the activity of malt diastase, Biochem. Zeitschr., 1916, vol. 77, pp. 146–167, quoted from Physiol. Abst., 1917, vol. 2, p. 17.

[279] Ambard, M., and Foà, C., Les modifications de l'acidité d'un mélange suc gastrique-albumine au cours de la digestion, C.-R. Soc. Biol., 1915, vol. 59, pp. 5–9.

[280] Auerbach, F., and Pick, H., Bemerkung zur Pankreasverdauung, Biochem. Zeitschr., 1913, vol. 48, pp. 425–426.

[281] Compton, A., The influence of the hydrogen concentration upon the optimum temperature of a ferment, Proc. Roy. Soc., London, (B) 1914, vol. 88, pp. 408–417.

[282] Van Dam., W., Beitrag zur Kenntnis der Labgerinnung, Zeitschr. f. physiol. Chem., 1909, vol. 58, pp. 295–330.

[283] Davidsohn, H., Die Pepsinverdauung im Säuglingsmagen unter Berücksichtigung der Acidität, Zeitschr. f. Kinderheilk., 1912, vol. 4, pp. 208–230.

[284] Davidsohn, H., Beitrag zum Studium der Magenlipase, Biochem. Zeitschr., 1912, vol. 45, pp. 284–302.

[285] Davidsohn, H., Ueber die Abhängigkeit der Lipase von der Wasserstoffionenkonzentration, Biochem. Zeitschr., 1913, vol. 49, pp. 249–277.

[286] Dernby, K. G., Étude sur la cinétique d'une hydrolyse enzymatique de la glycylglycocolle, C.-R. Lab., Carlsberg, 1916, vol. 11, pp. 263–294.

[287] Emslander, F., Die Wasserstoffionen-Konzentration im Biere und bei dessen Bereitung, Kolloid Zeitschr., 1913, vol. 13, pp. 156–169; 1914, vol. 14, pp. 44–48.

[288] Hägglund, E., Hefe und Gärung in ihrer Abhängigkeit von Wasserstoff- und Hydroxylionen, Sammlung chem. u. chem.-tech. Vorträge, 1914, vol. 21, pp. 129–174.

[289] Hägglund, E, Ueber die gärungshemmende Wirkung der Wasserstoffionen, Biochem. Zeitschr., 1915, vol. 69, pp. 181–191.

[290] Leberle, H., and Lüers, H., Acid determination in beer by electrometrical methods, Zeitschr. ges. Brau., vol. 37, pp. 177–184, quoted from Chemical Abst., 1914, vol. 8, p. 2447.

[291] Lüers, H., Change of hydrogen ion concentration during fermentation, Zeitschr. ges. Brau., vol. 37, pp. 79–82, quoted from Chemical Abst., 1914, vol. 8, p. 1845.

[292] Michaelis, L., Zur Theorie der elektrolytischen Dissoziation der Fermente, Biochem. Zeitschr., 1914, vol. 60, pp. 91–96.

[293] Michaels, L., and Davidsohn, H., Die isoelektrische Konstante des Pepsins, Biochem. Zeitschr., 1910, vol. 28, pp. 1–6.

[294] Michaelis, L., and Davidsohn, H., Trypsin und Pankreasnucleoproteid, Biochem. Zeitschr., 1910, vol. 30, pp. 481–504.

[295] Michaelis, L., and Davidsohn, H., Die Wirkung der Wasserstoffionen auf das Invertin, Biochem. Zeitschr., 1911, vol. 35, pp. 386–412.

[296] Michaelis, L., and Davidsohn, H., Die Abhängigkeit der Trypsinwirkung von der Wasserstoffionenkonzentration, Biochem. Zeitschr., 1911, vol. 36, pp. 280–290.

[297] Michaelis, L., and Mendelssohn, A., Die Wirkungsbedingungen des Pepsins, Biochem. Zeitschr., 1914, vol. 65, pp. 1–15.

[298] Michaelis, L., and Menten, M. L., Die Kinetik der Invertinwirkung, Biochem. Zeitschr., 1913, vol. 49, pp. 333–369.

[299] Michaelis, L., and Pechstein, H., Untersuchungen über die Katalase der Leber, Biochem. Zeitschr., 1913, vol. 53, pp. 320–355.

[300] Michaelis, L., and Pechstein, H., Die Wirkungsbedingungen der Speicheldiastase, Biochem. Zeitschr., 1914, vol. 59, pp. 77–99.

[301] Michaelis, L., and Rona, P., Ueber die Umlagerung der Glucose bei alkalischer Reaktion; ein Beitrag zur Theorie der Katalyse, Biochem. Zeitschr., 1912, vol. 47, pp. 447–461.

[302] Michaelis, L., and Rona, P., Die Wirkungsbedingungen der Maltase aus Bierhefe I. Biochem. Zeitschr., 1913, vol. 57, pp. 70–83.

[303] Morse, M., Hydrogen ion concentration in autolysis, Jour. Biol. Chem., 1916, vol. 24, Proc. Soc. Biol. Chem., p. xxvii.

[304] Norris, R. V., The hydrolysis of glycogen by diastatic enzymes. Comparison of preparations of glycogen from different sources, Biochem. Jour., 1913, vol. 7, pp. 26–42.

[305] Palitzsch, S., and Walbum, L. E., Sur la concentration optimale des ions hydrogène pour la première phase de la décomposition trypsique de la gélatine (Liquéfaction de la gelatine), C.-R. Lab., Carlsberg, 1912, vol. 9, pp. 200–236.

[306] Palitzsch, S., and Walbum, L. E., Ueber die optimale Wasserstoffionenkonzentration bei der tryptischen Gelatineverflüssigung, Biochem. Zeitschr., 1912, vol. 47, pp. 1–35.

[307] Reed, G. B., The relation of oxidase reactions to changes in hydrogen ion concentration, Jour. Biol. Chem., 1916, vol. 27, pp. 299–302.

[308] Reed, G. B., Measurement of oxidation potential, and its significance in the study of oxidases, Bot. Gaz., 1917, vol. 61, pp. 523–527.

[309] Ringer, W. E., and Van Tright, H., Einfluss der Reaktion auf die Ptyalinwirkung, Zeitschr. f. physiol. Chem., 1912, vol. 82, pp. 484–501.

[310] Robertson, T. B., and Schmidt, C. L. A., On the part played by the alkali in the hydrolysis of proteins by trypsin, Jour. Biol. Chem., 1908, vol. 5, pp. 31–48.

[311] Rohonyi, H., Die Veränderung der Wasserstoffionenkonzentration der Pepsinwirkung und das Säurebindungsvermögen einiger hydrolytischer Spaltungsprodukte des Eiweisses, Biochem. Zeitschr., 1912, vol. 44, pp. 165–179.

[312] Rona, P., Zur Kenntnis der Esterspaltung im Blute, Biochem. Zeitschr., 1911, vol. 33, pp. 413–422.

[313] Rona, P., and Arnheim, F., Beitrag zur Kenntnis des Erepsins, Biochem. Zeitschr., 1913, vol. 57, pp. 84–94.

[314] Rona, P., and Bien, Z., Zur Kenntnis der Esterase des Blutes, Biochem. Zeitschr., 1914, vol. 59, pp. 100–112.

[315] Rona, P., and Bien, Z., Vergleichende Untersuchungen über Pankreaslipase und Blutesterase, Biochem. Zeitschr., 1914, vol. 64, pp. 13–29.

[316] Rona, P., and Michaelis, L., Ueber Ester- und Fettspaltung im Blute und im Serum, Biochem. Zeitschr., 1911, vol. 31, pp. 345–354.

[317] Rona, P., and Michaelis, L., Die Wirkungsbedingungen der Maltase aus Bierhefe. II, Die Wirkung der Maltase auf a Methylglucosid und die Affinitätsgrösse des Ferments, Biochem. Zeitschr., 1913, vol. 58, pp. 148–157.

[318] Rona, P., and Wilenko, G. G., Beiträge zur Frage der Glykolyse IV, Biochem. Zeitschr., 1914, vol. 62, pp. 1–10.

[319] Sörensen, S. P. L., Enzymstudien. II, Ueber die Messung und die Bedeutung der Wasserstoffionenkonzentration bei enzymatischen Prozessen, Biochem. Zeitschr., 1909, vol. 21, pp. 131–304.

[320] Sörensen, S. P. L., Ergänzung zu der Abhandlung: Enzymstudien. II, Ueber die Messung und die Bedeutung der Wasserstoffionenkonzentration bei enzymatischen Prozessen, Biochem. Zeitschr., 1909, vol. 22, pp. 352–356.

[321] Sörensen, S. P. L., Études enzymatiques. II, Sur la mesure et l'importance de la concentration des ions hydrogène dans les reactions enzymatiques. C.–R. Lab., Carlsberg, 1909, vol. 8, pp. 1–168, 396–401.

[322] Van Slyke, D. D., and Zacharias, G., The effect of hydrogen ion concentration and of inhibiting substances on urease, Jour. Biol. Chem., 1914, vol. 19, pp. 181–210.

G. SEA WATER.

[323] Haas, A. R. C., The effect of the addition of alkali to sea water upon the hydrogen ion concentration, Jour. Biol. Chem., 1916, vol. 26, pp. 515–517.

[324] Henderson, L. J., and Cohn, E. J., Equilibrium between acids and bases in sea water, Proc. Nat. Acad. Sci., 1916, vol. 2, pp. 618–622.

[325] McClendon, J. F., New buffer mixtures, standard tubes, and colorimeter for determining the hydrogen ion concentration of sea water, Jour. Biol. Chem., 1916, vol. 28, Proc. Soc. Biol. Chem., pp. xxx–xxxi.

[326] Palitzsch, S., Ueber die Messung und die Grösse der Wasserstoffionenkonzentration des Meerwassers, Biochem. Zeitschr., 1911, vol. 37, pp. 116–130.

[327] Palitzsch, S., Sur le mésurage, et la grandeur de la concentration en ions hydrogène de l'eau salée, C.–R. Lab., Carlsberg, 1911, vol. 10, pp. 85–98.

[328] Sörensen, S. P. L., and Palitzsch, S., Sur "l'erreur de sel" dans la mesure colorimétrique de la concentration des ions hydrogène de l'eau de mer, C.–R. Lab., Carlsberg, 1911, vol. 10, pp. 252–258.

[329] Sörensen, S. P. L., and Palitzsch, S., Sur le mesurage de la concentration en ions hydrogène de l'eau de mer, C.–R. Lab., Carlsberg, 1910, vol. 9, pp. 8–37.

[330] Sörensen, S. P. L., and Palitzsch, S., Ueber die Messung der Wasserstoffionenkonzentration des Meerswassers, Biochem. Zeitschr., 1910, vol. 24, pp. 387–415.

III. BACTERIOLOGICAL.

[331] Ayers, S. H., Hydrogen-ion concentrations in cultures of streptococci, Jour. Bact., 1916, vol. 1, pp. 84–85.

[332] Beniasch, M., Die Säureagglutination der Bakterien, Zeitschr. f. Immunitätsfrsch. u. exper. Therap., 1912, vol. 12, pp. 268–315.

[333] Clark, W. M., The influence of hydrogen-ion concentrations upon the physiological activities of *Bacillus coli,* Science, n.s., 1915, vol. 41, p. 624.

[334] Clark, W. M., The final hydrogen ion concentrations of cultures of *Bacillus coli,* Jour. Biol. Chem., 1915, vol. 22, pp. 87–98.

[335] Clark, W. M., The "reaction" of bacteriologic culture media, Jour. Infec. Dis., 1915, vol. 17, pp. 109–136.

[336] Clark, W. M., and Lubs, H. A., The differentiation of bacteria of the colon-aerogenes family by the use of indicators, Jour. Infec. Dis., 1915, vol. 17, pp. 160–173.

[337] Clark, W. M., and Lubs, H. A., The colorimetric determination of hydrogen ion concentration and its applications in bacteriology, Jour. Bact., 1917, vol. 2, pp. 1–34, 109–136, 191–236.

[338] Gillespie, L. J., The acid agglutination of pneumococci, Jour. Exper. Med., 1914, vol. 19, pp. 28–37.

[339] Grote, L. R., Ueber die praktische Verwertbarkeit der Säureagglutination nach Michaelis, Centrbl. Bakt. Orig., 1913, vol. 69, pp. 98–104.

[340] Heimann, W., Die "Säureagglutination" innerhalb der Typhus-Paratyphusgruppe, insbesondere sogenannter Paratyphus- C- Baccillen, Zeitschr. f. Immunitätsfrsch., 1912, vol. 16, pp. 127–140.

[341] Henderson, L. J., and Webster, H. B., The preservation of neutrality in culture media with the aid of phosphates, Jour. Med. Res., 1907, vol. 16, pp. 1–5.

[342] Homer, A., The reaction of sera as a factor in the successful concentration of antitoxic sera by the methods at present in use, Biochem. Jour., 1917, vol. 11, pp. 21–39.

[343] Hurwitz, S. H., Meyer, K. F., and Ostenberg, Z., On a colorimetric method of adjusting bacteriological culture media to any optimum hydrogen ion concentration, Proc. Soc. Exper. Biol. and Med., 1915, vol. 13, pp. 24–26.

[344] Hurwitz, S. H., Meyer, K. F., and Ostenberg, Z., A colorimetric method for the determination of the hydrogen ion concentration of biological fluids, with special reference to the adjustment of bacteriological culture media, Johns Hopkins Hosp. Bull., 1916, 27, pp. 16–24.

[345] Itano, A., The relation of hydrogen ion concentration of media to the proteolytic activity of *Bacillus subtilis,* Mass. Agr. Exp. Sta. 1916, Bull. no. 167, pp. 139–177.

[346] Lindenschatt, S. M., Ueber den Einfluss der OH^- and H^+ ionen auf die Komplementablenkung und das differente Verhalten verschieden hoch erhitzter Sera bei der Komplementfixierung. Dissert. Heidelberg, 1913, pp. 1–39, quoted from Zentrbl. f. Biochem. u. Biophysik., 1914, vol. 16, p. 504.

[347] Markl, J. G., Ueber Säureagglutination von Pestbacillen, Centrbl. Bakt. Orig., 1915, vol. 77, pp. 102–108.

[348] Meyer, K., Zur Kenntniss der Bakterienproteasen, Biochem. Zeitsch., 1911, vol. 32, pp. 274–279.

[349] Michaelis, L., Die Säureagglutination der Bakterien, insbesondere der Typhusbazillen, Deut. med. Woch., 1911, vol. 37, pp. 969–971.

[350] Michaelis, L., and Marcora, F., Die Säureproduktivität des *Bacterium coli,* Zeitschr. f. Immunitätsfrsch. u. exper. Therap., 1912, vol. 14, pp. 170–173.

[351] Michaelis, L., and Skwirsky, P., Der Einfluss der Reaktion auf die spezifische Hämolyse, Zeitschr. f. Immunitätsfrsch. u. exper. Therap,. 1910, vol. 4, pp. 357–374, 629–635.

[352] Michaelis, L., and Takahashi, D., Die isoelektrischen Konstanten der Blutkörperchenbestandteile und ihre Beziehungen zur Säurehämolyse, Biochem. Zeitschr., 1910, vol. 29, pp. 439–452.

[353] Poppe, Dr., Die Säureagglutination der Bakterien der Paratyphusgruppe, Zeitschr. f. Immunitätsfrsch. u. exper. Therap., 1912, vol. 13, pp. 185–191.

[354] Walbum, L. E., Die Bedeutung der Wasserstoffionenkonzentration für die Hämolyse, Biochem. Zeitschr., 1914, vol. 63, pp. 221–268.

[355] Waterman, H. J., Ueber einige Faktoren welche die Entwickelung von *Penicillium glaucum* beeinflussen. Beitrag zur Kenntnis der Antiseptica und der Narkose, Centrbl. f. Bakt., 2. Abt., 1915, vol. 42, pp. 639–688.

IV. *S*OIL AND *P*LANT.

[356] Conner, S. D., Acid soils and effect of acid phosphate and other fertilizers upon them, Jour. Ind. Eng. Chem., 1916, vol. 8, pp. 35–40.

[357] Fischer, G., Die Säuren und Kolloide des Humus, Kühn Arch., 1914, vol. 4, pp. 1–36.

[358] Gillespie, L. J., The reaction of soil and measurements of hydrogen-ion concentration, Jour. Wash. Acad. Sci., 1916, vol. 6, pp. 7–16.

[359] Gillespie, L. J., and Hurst, L. A., Hydrogen ion concentration measurements of soils of two types: caribou loam and washburn loam, Soil Science, 1917, vol. 4, pp. 313–319.

[360] Haas, A. R. C., The acidity of plant cells as shown by natural indicators, Jour. Biol. Chem., 1916, vol. 27, pp. 233–241.

[361] Haas, A. R. C., Reaction of plant protoplasm, Bot. Gaz., 1917, vol. 63, pp. 225–228.

[362] Haas, A. R. C., The excretion of acid by roots, Proc. Nat. Acad. Sci., 1916, vol. 2, pp. 561–566.

[363] Haas, A. R. C., Anesthesia and respiration, Science, n.s., 1917, vol. 46, pp. 462–464.

[364] Hoagland, D. R., The effect of hydrogen and hydroxyl ion concentration on the growth of barley seedlings, Soil Science, 1917, vol. 3, pp. 547–560.

[365] Hoagland, D. R., and Sharp, L. T., Relation of carbon dioxide to soil reaction as measured by the hydrogen electrode, Jour. Agr. Res., 1918, vol. 12, pp. 139–148.

[366] Miyake, K., Toxic action of soluble aluminum salts upon the growth of the rice plant, Jour. Biol. Chem., 1916, vol. 25, pp. 23–28.

[367] Pantanelli, E., Ueber Ionenaufnahme, Jahrb. wiss. Bot., Pringsheim, 1915, vol. 56, pp. 689–733.

[368] Plummer, J. K., Studies in soil reaction as indicated by the hydrogen electrode, Jour. Agric. Res., 1918, vol. 12, pp. 19–31.

[369] Saidel, T., Quantitative Untersuchungen über die Reaktion wässeriger Bodenauszüge, Bull. Sect. Sci. Acad. Roumaine, 1913, Ann. 2, pp. 38–44.

[370] Sharp, L. T., and Hoagland, D. R., Acidity and adsorption in soils as measured by the hydrogen electrode, Jour. Agric. Res., 1916, vol. 7, pp. 123–145.

[371] Wagner, R. J., Wasserstoffionenkonzentration und natürliche Immunität der Pflanzen, Centrbl. f. Bakt., 2. Abt., 1916, vol. 44, pp. 708–719.

V. MISCELLANEOUS.

[372] Auerbach, F., and Pick, H., Die Alkalität wässeriger Lösungen kohlensäurer Salze, Arbeit. a. d. kais. Gesundheitsamte, 1912, vol. 38, pp. 243–274.

[373] Bethe, A., Die Bedeutung der Elektrolyten für die rhythmischen Bewegungen der Medusen, Angriffspunkt der Salze, Einfluss der Anionen und Wirkung der OH^- und H^+ Ionen, Arch. f. d. ges. Physiol., 1909, vol. 127, pp. 219–273.

[374] Bjerrum, N., Studien über chromichlorid, Zeitschr. f. phys. Chem., 1907, vol. 59, pp. 336–383.

[375] Bjerrum, N., Studien über chromichlorid. III, Hydroxoaquochromichloride, Zeitschr. f. phys. Chem., 1910, vol. 73, pp. 724–759.

[376] Brode, J., and Lange, W., Beiträge zur Chemie des Essigs mit besonderer Berücksichtigung seiner Untersuchungsverfahren, Arbeit. a. d. kais. Gesundheitsamte, 1909, vol. 30, pp. 1–54.

[377] Cohn, E. J., Relation between the hydrogen-ion concentration of sperm suspensions and their fertilizing power, Anat. Rec., 1917, vol. 11, p. 530.

[378] Dale, D., and Thacker, C. R. A., Hydrogen ion concentrations limiting automaticity in different regions of the frog's heart, Jour. Physiol., 1914, vol. 47, pp. 493–508.

[379] Hildebrand, J. H., and Bowers, W. G., A study of the action of alkali on certain zinc salts by means of the hydrogen electrode, Jour. Am. Chem. Soc., 1916, vol. 38, pp. 785–788.

[380] Hildebrand, J. H., and Harned, H. S., The rapid determination of magnesia in limestone by means of the hydrogen electrode, 8th Intern. Cong. Appl. Chem., 1912, vol. 1, pp. 217–225.

[381] Jessen-Hansen, H., Influence de la concentration en ions hydrogène sur la valeur boulangère de la farine, C.-R. Lab., Carlsberg., 1911, vol. 10, pp. 170–206.

[382] Loeb, J., Ueber die Ursachen der Giftigkeit einer reinen Chlornatriumlösung und ihrer Entgiftung durch K und Ca, Biochem. Zeitschr., 1906, vol. 2, pp. 81–110.

[383] Loeb, J., and Wasteneys, H., Die Beeinflüssung der Entwickelung und der Oxydationsvorgänge im Seeigelei (Arbacia) durch Basen, Biochem. Zeitschr., 1911, vol. 37, pp. 410–423.

[384] Michaelis, L., Die Säure- Dissoziationskonstanten der Alkohole und Zucker, insbeesondere der Methyl-glucoside, Ber. d. deut. chem. Ges., 1913, vol. 46, pp. 3683–3693.

[385] Michaelis, L., and Rona, P., Die Alkaliempfindlichkeit des Traubenzuckers, Biochem. Zeitschr., 1910, vol. 23, pp. 364–369.

[386] Paul, T., Der Säuregrad des Weines, Zeitschr. f. Electrochem., 1915, vol. 21, pp. 80–89.

[387] Sand, H. J. S., and Law, D. J., The employment of the electrometric method for the estimation of the acidity of tan liquors, Jour. Soc. Chem. Ind., 1911, vol. 30, pp. 3–5.

[388] Szili, A., Experimentelle Untersuchungen über Säureintoxikation, Arch. f. d. ges. Physiol., 1906, vol. 115, pp. 82–105.

[389] Szili, A., Weitere Untersuchungen über Vergiftung mit anorganischen und organischen Säuren, Arch. f. d. ges. Physiol., 1909, vol. 130, pp. 134–155.

[390] Wahl, R., New scientific conceptions and their application to quality and methods of preparing beer, Am. Brewers' Rev., 1915, vol. 29, pp. 271–274, 365–368, 557–559, quoted from Chemical Abst., 1916, vol. 10, p. 1398.

[391] Walker, J., and Kay, S. A., The acidity and alkalinity of natural waters, Jour. Soc. Chem. Ind., 1912, vol. 31, pp. 1013–1016.

[392] Wood, J. T., Sand, H. J. S., and Law, D. J., The employment of the electrometric method for the estimation of the acidity of tan liquors, Jour. Soc. Chem. Ind., 1911, vol. 30, pp. 872–877.

VI. Indicators.

[393] Acree, S. F., On the theory of indicators and the reactions of phthaleins and their salts, Am. Chem. Jour., 1908, vol. 39, pp. 528–544.

[394] Acree, S. F., and Slagle, E. A., On the theory of indicators and the reactions of phthaleins and their salts, Am. Chem. Jour., 1909, vol. 42, pp. 115–147.

[395] Bogert, M. T., and Scatchard, G., Researches on quinazolines xxxiii. A new and sensitive indicator for acidimetry and alkalimetry, and for the determination of hydrogen ion concentrations between the limits of 6 and 8 on the Sörensen scale, Jour. Am. Chem. Soc., 1916, vol. 38, pp. 1606–1615.

[396] Crozier, W. J., Some indicators from animal tissues, Jour. Biol. Chem., 1916, vol. 24, pp. 443–445.

[397] Fels, B., Studien über die Indikatoren der Acidimetrie und Alkalimetrie II, Zeitschr. f. Electrochem., 1904, vol. 10, pp. 208–214.

[398] Friedenthal, H., Die Bestimmung der Reaktion einer Flüssigkeit mit Hilfe von Indikatoren, Zeitschr. f. Electrochem., 1904, vol. 10, pp. 113–119.

[399] Harvey, E. N., A criticism of the indicator method of determining cell permeability for alkalies, Am. Jour. Physiol., 1913, vol. 31, pp. 335–342.

[400] Henderson, L. J., and Forbes, A., On the estimation of the intensity of acidity and alkalinity with dinitrohydroquinone, Jour. Am. Chem. Soc., 1910, vol. 32, pp. 687–689.

[401] Hottinger, R., Ueber "Lackmosol," den empfindlichen Bestandtheil des Indicators Lackmoid. Darstellung und einige Eigenschaften, Biochem. Zeitschr., 1914, vol. 65, pp. 177–188.

[402] Kelly, T. H., Hydrogen-ion acidity, Jour. Lab. Clin. Med., 1915, vol. 1, pp. 194–196.

[403] Lubs, H. A., and Acree, S. F., On the sulfonphthalein series of indicators and the quinone-phenolate theory, Jour. Am. Chem. Soc., 1916, vol. 38, pp. 2772–2784.

[404] Lubs, H. A., and Clark, W. M., On some new indicators for the colorimetric determination of hydrogen-ion concentration, Jour. Wash. Acad. Sci., 1915, vol. 5, pp. 609–617.

[405] Lubs, H. A., and Clark, W. M., A note on the sulphonephthaleins as indicators for the colorimetric determination of hydrogen-ion concentration, Jour. Wash. Acad. Sci., 1916, vol. 6, pp. 481–489.

[406] Michaelis, L., and Rona, P., Zur Frage der Bestimmung der H-Ionenkonzentration durch Indikatoren, Zeitschr. f. Electrochem., 1908, vol. 14, pp. 251–253.

[407] Michaelis, L., and Rona, P., Der Einfluss der Neutralsalze auf die Indicatoren, Biochem. Zeitschr., 1909, vol. 23, pp. 61–67.

[408] Noyes, A. A., Quantitative application of the theory of indicators to volumetric analysis, Jour. Am. Chem. Soc., 1910, vol. 32, pp. 815–861.

[409] Palitzsch, S., Sur l'emploi du rouge de méthyle au mesurage colorimétrique de la concentration en ions hydrogène, C.-R. Lab., Carlsberg, 1911, vol. 10, pp. 162–169.

[410] Palitzsch, S., Ueber die Verwendung von Methylrot bei der colorimetrischen Messung der Wasserstoffionenkonzentration, Biochem. Zeitschr., 1911, vol. 37, pp. 131–138.

[411] Rosenstein, L., The ionization constant of phenolphthalein and the effect upon it of neutral salts, Jour. Am. Chem. Soc., 1912, vol. 34, pp. 1117–1128.

[412] Salessky, W., Ueber Indikatoren der Acidimetrie und Alkalimetrie I, Zeitschr. f. Electrochem., 1904, vol. 10, pp. 204–208.

[413] Salm, E., Die Bestimmung des H^+-Gehaltes einer Lösung mit Hilfe von Indikatoren, Zeitschr. f. Electrochem., 1904, vol. 10, pp. 341–346.

[414] Salm, E., Kolorimetrische Affinitätsmessungen, Zeitschr. f Electrochem., 1906, vol. 12, pp. 99–101.

[415] Salm, E., Studie über Indikatoren, Zeitschr. f. phys. Chem., 1906, vol. 57, pp. 471–501.

[416] Salm, E., Messungen der Affinitätsgrössen organischer Säuren mit Hilfe von Indikatoren, Zeitschr. f. phys. Chem., 1908, vol. 63, pp. 83–108.

[417] Salm, E., and Friedenthal, H., Zur Kenntnis der acidimetrischen und alkalimetrischen Indikatoren, Zeitschr. f. Electrochem., 1907, vol. 13, pp. 125–130.

[418] Scatchard, G., and Bogert, M. T., A new and very sensitive indicator for acidimetry and alkalimetry and for determining hydrogen ion concentrations between the limits of 6 and 8 on the Sörensen scale, Science, n.s., 1916, vol. 43, p. 722.

[419] Sörensen, S. P. L., and Palitzsch, S., Sur un indicateur nouveau, α-naphtolphtaléine, ayant un virage au voisinage du point neutre, C.-R. Lab., Carlsberg, 1910, vol. 9, pp. 1–7.

[420] Sörensen, S. P. L., and Palitzsch, S., Ueber den "Salzfehler" bei der colorimetrischen Messung der Wasserstoffionenkonzentration des Meerwassers, Biochem. Zeitsch., 1913, vol. 51, pp. 307–313.

[421] Stieglitz, J., The theories of indicators, Jour. Am. Chem. Soc., 1903, vol. 25, pp. 1112–1127.

[422] Thiel, A., Der Stand der Indikatorenfrage, Sammlung chem. u. chem.-tech. Vorträge, 1911, vol. 16, pp. 307–422.

[423] Tizard, H. T., The colour changes of methyl-orange and methyl-red in acid solution, Jour. Chem. Soc., London, 1910, vol. 97, pp. 2477–2490.

[424] Tizard, H. T., The hydrolysis of aniline salts measured colorimetrically, Jour. Chem. Soc., London, 1910, vol. 97, pp. 2490–2495.

[425] Walbum, L. E., Ueber die Verwendung von Rotkohlauszug als Indikator bei der colorimetrischen Messung der Wasserstoffionenkonzentration, Biochem. Zeitschr., 1913, vol. 48, pp. 291–296.

[426] Walbum, L. E., Sur l'emploi de l'extrait de choux rouge comme indicateur dans la mesure colorimétrique de la concentration des ions hydrogène, C.–R. Lab., Carlsberg, 1911, vol. 10, pp. 227–232.

[427] Walpole, G. S., Chart presentation on recent work on indicators, Biochem. Jour., 1910, vol. 5, pp. 207–214.

[428] Wegscheider, R., Ueber den Farbenumschlag des phenolphthaleins, Zeitschr. f. Electrochem., 1908, vol. 14, pp. 510–512.

ADDENDA

[429] Baker, J. C., and Van Slyke, L. L., A method for making electrometric titrations of solutions containing protein, Jour. Biol. Chem., 1918, vol. 35, pp. 137–145.

[430] Barnett, G. D., and Chapman, H. S., Colorimetric determination of reaction of bacteriologic mediums and other fluids, Jour. Amer. Med. Assoc., 1918, vol. 70, pp. 1062–1063.

[431] de Corral y Garcia, J. M., La reaccion actual de la sangre y su determinacion electrometrica. Tesis de doctorado, Universidad de Madrid, Valladolid, 1914, pp. 1–162.

[432] Crozier, W. J., On indicators in animal tissues, Jour. Biol. Chem., 1918, vol. 35, pp. 455–460.

[433] Cullen, G. E., and Austin, J. H., Hydrogen ion concentrations of various indicator end-points in dilute hydochlorite solutions, Jour. Biol. Chem., 1918, vol. 34, pp. 553–568.

[434] Evans, A. C., A study of the streptococci concerned in cheese ripening, Jour. Agri. Res., 1918, vol. 13, pp. 235–252.

[435] Evans, A. C., Bacterial flora of Roquefort cheese, Jour. Agric. Res., 1917, vol. 13, pp. 225–233.

[436] Fred, E. B., and Loomis, N. E., The influence of hydrogen-ion concentration, of the medium on the reproduction of alfalfa bacteria. Jour. Bact., 1917, vol. 2, pp. 629–633.

[437] Gainey, P. L., Soil reaction and the growth of azotobacter, Jour. Agric. Res., 1918, vol. 14, pp. 265–271.

[438] Gainey, P. L., Soil reaction and the presence of azotobacter, Science, n.s., 1918, vol. 48, pp. 139–140.

[439] Gillespie, L. J., Correlation of H-ion exponent and occurrence of bacteria in soil, Science, n.s., 1918, vol. 48, pp. 393–394.

[440] Gillespie, L. J., and Wise, L. E., Action of neutral salts on humus and other experiments on soil acidity, Jour. Amer. Chem. Soc., 1918, vol. 40, pp. 796–813.

[441] Gillespie, L. J., The growth of the potato scab organism at various hydrogen ion concentrations as related to the comparative freedom of acid soils from the potato scab, Phytopathology, 1918, vol. 8, pp. 257–269.

[442] Goldberger, J., The change in the hydrogen-ion concentration of muscle during work, Biochem. Zeitschr., 1917, vol. 84, pp. 201–209, after Chem. Abs., 1918, vol. 12, p. 1482.

[443] Haas, A. R. C., On the preparation of ovalbumin and its refractive indices in solution, Jour. Biol. Chem., 1918, vol. 35, pp. 119–125.

[444] Hasselbalch, K. A., The reduced and the regulated "hydrogen figure" of the blood, Biochem. Zeitschr., 1916, vol. 74, pp. 56–62, via Physiol. Abs., 1916, vol. 1, p. 252.

[445] Hoagland, D. R., Studies on the relation of the nutrient solution to the composition and reaction of the cell sap of the barley plant, Botanical Gazette (in press).

[446] Hoagland, D. R., The relation of the plant to the reaction of the nutrient solution, Science, n.s., 1918, vol. 48, pp. 422–425.

[447] Hoagland, D. R., and Christie, A. W., The chemical effects of CaO and $CaCo_3$ on the soil, Soil Science, 1918, vol. 5, pp. 379–382.

[448] Höber, R., Die Gaskettenmethode zur Bestimmung der Blutreaktion, Deut. med. Woch., 1917, vol. 43, pp. 551–552, via Physiol. Abs., 1918, vol. 2, p. 604.

[449] Homer, A., A note on the use of indicators for the colorimetric determination of the hydrogen-ion concentration of sera, Biochem. Jour., 1917, vol. 11, pp. 283–291.

[450] Kligler, I. J., The effect of hydrogen-ion concentration on the production of precipitates in a solution of peptone and its relation to the nutritive value of media, Jour. Bact., 1917, vol. 2, pp. 351–353.

[451] Meacham, M. R., The hydrogen-ion concentration necessary to inhibit growth of four wood destroying insects, Science, n.s., 1918, vol. 48, pp. 499–500.

[452] Noyes, A. A., and Chow, M., The potentials of the bismuth-bismuthoxychloride and the copper-cuprouschloride electrodes, Jour. Amer. Chem. Soc., 1918, vol. 40, pp. 739–763.

[453] Osterhout, W. J. V., The determination of buffer effects in measuring respiration, Jour. Biol. Chem., 1918, vol. 35, pp. 237–240.

[454] Osterhout, W. J. V., A method of studying respiration. Jour. Gen. Phys., 1918, vol. 1, pp. 17–22.

[455] Osterhout, W. J. V., and Haas, A. R. C., On the dynamics of photo-synthesis, Jour. Gen. Physiol., 1918, vol. 1, pp. 1–16.

[456] Robertson, T. B., Ueber die verbindungen der Proteine mit anorganischen Substanzen und ihre Bedeutung für die Lebensvorgänge, Ergeb. der Physiol., 1910, vol. 10, pp. 216–361.

[457] Sherman, H. C., Thomas, A. W., and Baldwin, M. E., Influence of hydrogen-ion concentration upon the enzymic activity of three typical amylases, Proc. Soc. Exp. Biol. and Med., 1918, vol. 16, pp. 17–18.

[458] Truog, E., Soil acidity: I, Its relation to the growth of plants, Soil Science, 1918, vol. 5, pp. 169–195.

[459] Szili, A., The reaction of human milk, Biochem. Zeitschr., 1917, vol. 84, pp. 194–200, quoted after Chem. Abs., 1918, vol. 12, p. 1482.

[460] Van Slyke, L. L., and Baker, J. G., Free lactic acid in sour milk, Jour. Biol. Chem., 1918, vol. 35, pp. 147–178.

neumococci. Jour. Exp. Med., 1918, vol. 28, pp. 289–296.
rnby, K. G., and Avery, O. T., Optimum hydrogen-ion con-
for the growth of pneumococcus. Jour. Exp. Med., 1918,
p. 345–357.
lespie, L. J., and Hurst, L. A., Hydrogen-ion concentration
pe—common potato scab. Soil Science, 1918, vol. 6, pp.

ɔe, F. E., and Osugi, S., The inversion of cane sugar by soils
d substances and the nature of soil acidity, Soil Science,
5, pp. 333–358.
ole, E. H., and Tottingham, W. E., The influence of certain
ids upon the composition and efficiency of Knop's nutrient
Amer. Jour. Bot., 1918, vol. 5, pp. 452–461.
ıksman, S. A., The occurrence of azotobacter in cranberry
ənce, n.s., 1918, vol. 48, pp. 653–654.

bibliography has been brought up to date of publication.

be Stimulated Chemically? (Preliminary
. Maxwell. Pp. 17-19. February, 1906........ .05
ropism in Paramecium by Chemical Sub-
ancroft. Pp. 21-23. March, 1906........ .10
ric Oxygen for the Eggs of the Sea-urchin
purdtus) after the Process of Membrane
Loeb. Pp. 33-37. March, 1906.
Presence of Free Oxygen in the Hypertonic
duction of Artificial Parthenogenesis, by
7. March, 1906.
over........ .15
the Toxic Effect of Hypertonic Solutions
Unfertilized Egg of the Sea-urchin by Lack
Loeb. Pp. 49-56. April, 1906........ .05
Fertilization Membrane in the Egg of the
ood of Certain Gephyrean Worms (a pre-
es Loeb. Pp. 57-58. March, 1907........ .05
a Protein through the Action of Pepsin
ation), by T. Brailsford Robertson. Pp.
.... .05
f the Process of Fertilization, and its Bear-
f Life-Phenomena, by Jacques Loeb. Pp.
.... .25
neability of Cells for Salts or Ions (a pre-
), by Jacques Loeb. Pp. 81-86. January,
.... .05
Retrogressive Varieties by one Mutation in
doorn. Pp. 87-90. September, 1908........ .05
in through the Agency of Pepsin and Chemi-
lysis and Synthesis of Proteins through the
y T. B. Robertson. Pp. 91-94. December,
.... .05
Color in Rodents, by Arend L. Hagedoorn.
.... .05
ncentrations corresponding to Electromotive
s-chain measurements, by C. L. A. Schmidt.
r, 1909........ .10
ford Robertson. Pp. 115-194. October, 1910 $1.00
tity of Heliotropism in Animals and Plants,
. S. Maxwell. Pp. 195-197. January, 1910 .05
tion of the Internal Ear, by S. S. Maxwell.
10 .... .05
in Rabbits, Caused by the Injection of Salt
Burnett. Pp. 5-7. September, 1910 .... .05
of Dominance, by A. R. Moore. Pp. 9-15.
.... .05
in *Gonium pectorale*, by A. R. Moore and T.
3. May, 1911 .... .05
Biological Individuality of the Tissues and
Species, by T. Brailsford Robertson. Pp.
.... .05
the Quantitative Determination of Gliadin,
1-74. August, 1911 .... .40
rring in the Students of the University of
Burnett. Pp. 75-77. August, 1911 .... .05
of Ox-Blood Serum upon Sea-Urchin Eggs,
Proteins (*Preliminary communication*), by T.
Pp. 79-88. February, 1912 .... .10
tical Layer in Sea Urchin Eggs, by A. R.
ch, 1912.
nsitization of Sea Urchin Eggs by Strontium
re. Pp. 91-93. March, 1912.
cover........ .05
se, the Fertilizing and Cytolyzing Substance
ra, by T. Brailsford Robertson. Pp. 95-102.

UNI

12. Or
13. Th
14. A
15. Ca
16. On
17. Ne
18. On
19. The
20. A
21. Tal

Vol. 5. 1. The
2. The
3. The
4. Tal

Other series:
nomics, Engineeri
servatory Publica
Publications of th

CPSIA information can be obtained
at www.ICGtesting.com
Printed in the USA
BVHW091416141118
533117BV00013B/1014/P